LAB MANUAL
TO ACCOMPANY

ELECTRICITY FOR REFRIGERATION, HEATING, AND AIR CONDITIONING 8TH EDITION

Russell E. Smith

DELMAR
CENGAGE Learning

Australia • Brazil • Japan • Korea • Mexico • Singapore • Spain • United Kingdom • United States

DELMAR
CENGAGE Learning™

Lab Manual to Accompany Electricity for Refrigeration, Heating and Air Conditioning, 8e
Author(s): Russell E. Smith

Vice President, Career and Professional Editorial: Dave Garza

Director of Learning Solutions: Sandy Clark

Senior Acquisitions Editor: James DeVoe

Managing Editor: Larry Main

Senior Product Manager: John Fisher

Editorial Assistant: Cris Savino

Vice President, Career and Professional Marketing: Jennifer McAvey

Marketing Director: Deborah Yarnell

Marketing Manager: Katie Hall

Marketing Coordinator: Mark Pierro

Production Director: Wendy Troeger

Production Manager: Mark Bernard

Content Project Manager: David Plagenza

Senior Art Director: Casey Kirchmayer

Technology Project Manager: Joe Pliss

For product information and technology assistance, contact us at
Professional Group Cengage Learning Customer & Sales Support, 1-800-354-9706

For permission to use material from this text or product, submit all requests online at **www. cengage.com/permissions.** Further permissions questions can be e-mailed to **permissionrequest@cengage.com**

Library of Congress Control Number: 2010925357

ISBN-13: 978-1-111-03875-5
ISBN-10: 1-111-03875-9

Delmar
5 Maxwell Drive
Clifton Park, NY 12065-2919
USA

Cengage Learning is a leading provider of customized learning solutions with office locations around the globe, including Singapore, the United Kingdom, Australia, Mexico, Brazil and Japan. Locate your local office at: **international. cengage.com/region**

Cengage Learning products are represented in Canada by Nelson Education, Ltd.

For your lifelong learning solutions, visit **delmar.cengage.com.**

Visit our corporate website at **cengage.com.**

NOTICE TO THE READER
Publisher does not warrant or guarantee any of the products described herein or perform any independent analysis in connection with any of the product information contained herein. Publisher does not assume, and expressly disclaims, any obligation to obtain and include information other than that provided to it by the manufacturer. The reader is expressly warned to consider and adopt all safety precautions that might be indicated by the activities described herein and to avoid all potential hazards. By following the instructions contained herein, the reader willingly assumes all risks in connection with such instructions. The publisher makes no representations or warranties of any kind, including but not limited to, the warranties of fitness for particular purpose or merchantability, nor are any such representations implied with respect to the material set forth herein, and the publisher takes no responsibility with respect to such material. The publisher shall not be liable for any special, consequential, or exemplary damages resulting, in whole or part, from the readers' use of, or reliance upon, this material.

Printed in the United States
1 2 3 4 5 6 7 14 13 12 11 10

CONTENTS

PREFACE

This Lab Manual was prepared by the author as a supplement to *Electricity for Refrigeration, Heating, and Air Conditioning,* 8th Edition. The 18 chapters of the Lab Manual correlate with the same chapters in the 7th edition of the text. Each chapter is divided into the following sections to help students organize their study efforts and to help them master skills that they will be required to perform as refrigeration, heating, and air-conditioning technicians in the industry.
- Chapter Overview
- Key Terms
- Review Test
- Lab Exercises (not applicable to all chapters)

CHAPTER OVERVIEW

The **chapter overview** is a brief summary of the material covered in the text. The material provided in the chapter overview is not intended to provide the student with adequate information to successfully complete the learning activities and lab exercises for that chapter.

KEY TERMS

The **key terms** are taken from the same chapter in the text and considered to be extremely important to the student. No definitions are provided in the Lab Manual to encourage the student to define the terms from memory or to research the terms from the text or other available materials.

REVIEW TEST

The **review test** is provided as a means of determining if the student has an adequate understanding of the material in the corresponding chapter. The review test may be used as an end-of-chapter test or as a supplement before completing the lab exercises.

LAB EXERCISES

Lab exercises are provided for each chapter where practical. Every attempt has been made to make each lab exercise simulate the activities that are performed by a technician in the HVAC industry. Each lab exercise provides the following:
- Objectives
- Introduction

- Text Reference(s)
- Tools and Materials
- Safety Precautions (if applicable)
- Lab Sequences
- Maintenance of Work Station and Tools
- Summary Statements
- Questions

Objectives are written in a clear and precise manner, informing the student of the procedures and skills that will be gained from each lab exercise.

Introductions give the student reasons to become proficient at the activities in the particular lab exercise and provide a description of the lab activities.

Text references refer the student to the section in the text where the material is covered so that the student has easy access to review material that relates to each lab exercise.

Tools and materials that are needed to complete the lab exercises are listed, enabling the student to complete the lab in a timely and efficient manner.

Safety precautions are extremely important for the student when working in a laboratory environment. Safety procedures should be emphasized to the student in order to prevent accidents both in the lab and on the job.

Lab sequences provide step-by-step instructions for completing the exercise.

The maintenance of work station and tools is important to maintain a clean and safe working area both in the lab and on the job. Students are required to replace the covers on equipment, clean work areas, and keep up with tools in order to develop good work habits.

Summary statements are provided that are directly related to the lab exercise that the student has just completed.

Questions are provided to reinforce the activities completed in the lab exercise.

Each lab exercise was selected to enable the student to use the material covered previously in a classroom setting. Some of the lab exercises are designed to reinforce basic concepts, such as electrical circuits, electrical symbols, and the ability to read and draw schematic diagrams. However, most lab exercises are practical and cover skills that technicians are required to perform on the job. These lab exercises cover most components as a single device and expand to control systems and circuitry using these electrical devices. The lab exercises are practical and driven by tasks that technicians perform on average service calls. The lab exercises are designed so that most institutions with course offerings in heating, air conditioning, and refrigeration can easily utilize these lab exercises with equipment and supplies that are presently available in the program areas.

In these lab exercises, the student progresses from simple to advanced problems, both in reading schematic diagrams and in troubleshooting components and then troubleshooting systems. The program instructor will have to organize certain electrical component kits in order for the student to troubleshoot components that are faulty; the components used in these kits will have to be screened so that the student will see most faults that commonly occur in components in the industry. For example, in the relay kit, the instructor will have to place relays with bad normally open and closed contacts and with bad solenoid coils. There are times the instructor will have to place problems in equipment in order for the student to have the experience of troubleshooting control systems and components in heating, cooling, and refrigeration equipment. Instructors will have to maintain equipment in operable condition so that the student can install and operate it in the lab. The lab exercises are intended to offer students as many learning experiences as possible in the troubleshooting of components and equipment.

INSTRUCTOR'S RESOURCE CD

This educational resource creates a truly electronic classroom. It is a CD-ROM containing tools and instructional resources that enrich the classroom and make the instructor's preparation time shorter. The elements of the *Instructor's resource* link directly to the text to provide a unified instructional system.

With the *Instructor's resource* you can spend your time teaching, not preparing to teach. (ISBN 1111038767)

Features contained in the *Instructor's resource* include the following:

- **Instructor's Guide.** This PDF file contains a short description of the material covered in each chapter and answers to questions in the text and Lab Manual. Information regarding certain component kits and their composition, troubleshooting problems that the instructor will have to place in equipment, and components and equipment needed for lab exercises also can be found in the Instructor's Guide. The Instructor's Guide also provides suggestions pertaining to the organization of the material and general teaching tips.
- **Lesson Plans.** Each chapter has a lesson overview, objectives, key terms, and assignments provided.
- **ExamView Test Bank.** Over 300 questions of varying levels of difficulty are provided in true/false, fill-in-the-blank, and short answer formats so you can assess student comprehension. This versatile tool enables the instructor to manipulate the data to create original tests.
- **PowerPoint Presentations.** These slides provide the basis for a lecture outline to present concepts and material. Key points and concepts can be graphically highlighted for student retention.
- **Optical Image Gallery.** This database of key images taken from the text can be used in lecture presentations, tests and quizzes, and PowerPoint presentations. Additional Image Masters tie directly to the chapters and can be used in place of transparency masters.

ACKNOWLEDGMENTS

The author would like to thank the following reviewers for their input:

Greg Jourdan, Wenatchee Valley College

Joe Moravek, Lee College

Darius Spence, Northern Virginia Community College

Richard Wagoner, San Joaquin Valley College

Joel Warford, Vatterott College

CHAPTER 1 Electrical Safety

Chapter Overview

Electricity is commonplace in our environment today; in fact, it's hard for us to envision life without electricity. Examination of modern lifestyles indicates just how important electricity is to society. Society uses many small personal appliances that are considered a necessity, such as hair dryers, hair curlers, and toothbrushes. Mixers, toasters, and microwaves are common in almost every kitchen. Modern household living would be totally different without refrigerators, freezers, and electric stoves. Electricity plays an even larger part in maintaining comfortable conditions in residences, businesses, public buildings, and the workplace. No matter what a person does in our society, they are likely to come in contact with electrical power sources that are dangerous.

The most common injury related to electricity is electrical shock. An electrical shock occurs when a person becomes part of an electrical circuit. When electricity passes through the human body, the results can range from death to a slightly uncomfortable stinging sensation, depending upon the amount of electricity that passes through the body, the path that the electricity takes, and the duration that the electricity flows. The technician working on refrigeration, heating, or air conditioning in the field will find it impossible to install or troubleshoot air conditioning equipment without working close to electrical devices that are being supplied with electrical energy. The technician should develop a procedure to work around live electrical circuits without coming in contact with conductors and electrical components that are being supplied with electrical power. The technician should never allow part of his or her body to become part of an electrical circuit by coming in contact with a source of electrical power and the neutral or a ground. The path that current flow takes through the body is related to the severity of the electrical injury. If a technician allows the path of current flow to cross his or her heart, it could cause the heart to flutter rather than beat. This condition can be fatal unless a person close by knows cardiopulmonary resuscitation (CPR). Technicians can be burned by an electrical shock from higher voltages. Many technicians are injured when they are shocked and their reflex causes them to jerk away from the electrical source. Technicians should be aware of the possibility of ladders coming in contact with electrical sources. Technicians should be cautious and attentive when working near live electrical circuits.

Often when people receive an electrical shock, they will be unable to release the object that is the source of the electrical energy. People who are assisting the shock victim must make certain that they do not touch the victim unless the victim is no longer in contact with the electrical source. Once the rescuer has determined that the victim is no longer touching the electrical source, he or she can proceed to help. As soon as the shock victim is away from the electrical source, the rescuer should start first aid procedures. If the rescuer has determined that the victim's heart is not beating, then CPR should be started as soon as possible (within four to six minutes) or permanent damage may occur. It is advisable that one person on each service or installation truck be trained to perform CPR.

When installing electrical circuits in a structure, the technician must follow state and local codes along with the *National Electrical Code.*® (The *National Electrical Code*® and *NEC*® are registered trademarks of the National Fire Protection Association, Inc., Quincy, MA 02269.) All electrical circuits should be protected in the event of a circuit overload. All conductors should be large enough for the ampacity required by the circuit.

The ground wire is used in an electrical circuit to allow for a path of current flow back through the ground instead of through a person, thus causing electrical shock. If the path of an electrical conductor comes in contact with the metal frame of a piece of equipment that is not grounded, and a person touches the frame, he or she will become part of the circuit with possible bodily injury. In almost all cases, the ground wire can be identified by the color green. When working with power tools, make certain that they are double insulated or are equipped with a three-prong grounding plug. A ground fault receptacle should be used when working with power tools on a construction site to protect the operator.

When a technician is performing work on a circuit where there is a possibility that someone might accidentally restore electrical power to that circuit, the technician should place a padlock and warning label on the applicable switch or circuit breaker. If working in a residence, technicians must make the homeowner aware that they are working on the equipment and the homeowner should not turn the breaker or switch on.

Electricity cannot be seen, but certainly can be felt. It only takes a small amount of electricity to cause serious injury or death. It is imperative that heating and air-conditioning technicians respect and be cautious around electrical circuits. It only takes a slip or careless move for technicians to find themselves in danger of electrocution or injury. The technician must be careful and cautious around live electrical circuits. The technician is responsible for his or her own safety and should learn to respect and work carefully around electrical circuits.

Key Terms

Cardiopulmonary resuscitation (CPR)	**Electrical shock**	**Grounding adapter**
Circuit breaker	**Electromotive force**	**Live electrical circuit**
Circuit lockout	**Fuse**	*National Electrical Code*®
Conductor	**Ground**	**Three-prong plug**
Double insulated	**Ground fault circuit interrupter (GFCI)**	

REVIEW TEST

Name: _____   Date: _____   Grade: ____

Mark the following statements True or False.

1. _____ Air-conditioning technicians at times have to work close to live electrical circuits.

2. _____ Technicians should develop a safe procedure to work around live electric circuits.

3. _____ A technician's body can become part of an electrical circuit if he or she touches an air-conditioning unit that is properly grounded.

4. _____ A technician's body can become part of an electrical circuit if he or she comes in contact with the two bare conductors supplying power to a disconnect switch.

5. _____ There is no danger when electrical flow crosses the heart.

6. _____ The arc from a high-voltage transformer can cause burns on the body.

7. _____ It takes a high current flow through the body to prove fatal.

8. _____ Many times technician injuries come from the automatic reflex to get away from the electrical source when the technician is shocked.

9. _____ Technicians should always be aware of the location of power lines when they are trying to place a ladder on a structure.

10. _____ When assisting a shock victim, the rescuer should immediately give the victim CPR.

11. _____ The rescuer should make certain that a shock victim is not touching any source of electrical power before attempting any kind of rescue effort.

12. _____ The *National Electrical Code* specifies the minimum standards that must be met for safe installation of electrical systems.

13. _____ The ground wire used on air-conditioning equipment should connect the equipment to the power source.

14. _____ It is safe to use an electric drill if the grounding blade of a three-prong plug has been removed.

15. _____ All electrical circuits should be protected by following *National Electrical Code* standards.

16. _____ The technician can use a circuit breaker with a higher amp rating than necessary to prevent nuisance trip-outs.

17. _____ When a technician is repairing equipment in a location where a disconnect could be turned on, he or she should use proper circuit lockout procedures.

18. _____ The technician should examine all extension cords and power tools before using them.

19. _____ The technician should not bother to tell the homeowner that he or she had turned a circuit breaker off at the service panel.

20. _____ A ground fault circuit interrupter opens when a small electrical leak to ground is detected.

Answer the following questions.

1. What procedure should a technician follow when a fellow employee receives an electrical shock and is down?

2. Why is circuit grounding important?

3. Why is it important to follow the *National Electrical Code* and local and state codes?

4. Why should a technician lock out a circuit when working on equipment?

5. List six safety guidelines that should be used by air-conditioning technicians working on equipment in the industry.

CHAPTER 2 — Basic Electricity

Chapter Overview

Electricity has played and will continue to play an important part in the technological advancements of society. Electricity is converted to other forms of energy, such as light, heat, sound, and magnetism. In the air-conditioning industry, some of the more important uses of electrical energy are to produce heat through an electrical resistance heater and magnetism, which produces the magnetic field that allows relays and contactors to open and close, motors to rotate, and solenoid valves to open and close. The air-conditioning industry relies heavily on electricity for the operation of any heating, cooling, or refrigeration system.

A basic overview of atomic theory is the first step in understanding electricity. Every physical object is composed of matter, whether it is a solid, liquid, or gas. These elements are composed of atoms. An atom is the smallest particle of an element that can exist alone or in combination. All matter is made up of atoms or a combination of atoms (known as *molecules*), and all atoms are electrical in structure. A molecule is the smallest particle of a substance that has the properties of that substance. An atom is the smallest particle that can combine with other atoms to form molecules. Atoms are made up of very small particles. The nucleus is composed of protons and neutrons with electrons orbiting around the nucleus. Electrons are negatively charged particles while protons are positively charged particles. The neutron has a neutral charge that tends to hold the protons together in the nucleus.

An atom usually has an equal number of protons and electrons. When this condition exists, the atom is electrically neutral because the positively charged protons and negatively charged electrons are balanced. When an atom loses or gains an electron, it becomes unbalanced. The loss of an electron causes the atom to be positively charged and the addition of an electron causes the atom to be negatively charged. When one atom is charged and there is an unlike charge in another atom, electrons can flow between the two. This electron flow is called *electricity*. Like electrical charges repel and unlike charges attract.

Electricity can be produced by friction (static electricity), chemicals (battery), and magnetism (generator). Conductors allow electrons to easily flow from one atom to another. Insulators retard the flow of electrons from atom to atom.

Electrical potential is the force that moves electrons in an electric circuit much like the pressure in a water system that forces water through a pipe. The higher the pressure in a water circuit, the larger the flow of water. In an electrical circuit, the higher the electrical potential, the larger the current flow. "Voltage", "potential difference", and "electromotive force" are all terms used to describe electrical pressure. The flow of electrons in an electrical circuit is called *current*. Direct and alternating are two types of electrical current. Direct current flows in only one direction while alternating current alternates or reverses direction. The electric current in an electrical circuit is measured in amperes using an ammeter. Resistance is the opposition to current flow in an electrical circuit. Resistance is measured in ohms using an ohmmeter.

Electric power is the rate at which electrons do work or the rate at which electrons are being used. The power of an electric circuit is measured in watts. Common conversion factors used in the industry are 1 horsepower = 746 watts and 1 watt = 3.41 Btus per hour. A kilowatt-hour is the unit used by electrical utilities to bill customers for their energy usage.

Ohm's law ($E = IR$) states the mathematical relationship between current, electromotive force, and resistance in an electrical circuit. When any two factors in an electric circuit are known or can be measured, the formulas for Ohm's law can be used to find the third factor. Electric power can be calculated by using the formula $P = IE$.

Key Terms

Ampere	Electron	Nucleus
Atom	Element	Ohm
Compound	Field of force	Ohm's law
Conductors	Free electrons	Power factor
Current	Insulators	Proton
Electric energy	Kilowatt-hour	Seasonal energy efficiency ratio
Electric power	Law of electric charges	Static electricity
Electricity	Matter	Volt
Electrodes	Molecule	Voltage
Electrolytes	Neutron	Watt

REVIEW TEST

Name: _____   Date: _____   Grade: ___

Use the following to fill in the blanks.

alternating current	**Kilowatt-hour**
amperes	**matter**
atom	**molecule**
attract	**Ohm's law**
conductor	**potential difference**
current	**repel**
direct current	**resistance**
electric power	**SEER**
electrical pressure	**static electricity**
electromotive force	**voltage**
elements	**watt**
insulator	

1. _____ is the substance of which a physical object is composed.

2. Matter is composed of fundamental substances called _____.

3. The smallest particle of an element that can exist alone or in combination is a(n) _____.

4. The smallest particle of a substance that maintains the properties of that substance is a(n) _____.

5. The law of electric charges states that like charges _____ and unlike charges _____.

6. Friction is usually the cause of _____.

7. A material with free electrons is called a(n) _____.

8. A material that retards the flow of electrons is called a(n) _____.

9. _____, _____, and _____ are all terms used to describe electric pressure.

10. If an electrical circuit has a complete path and _____ is applied, then electrons will flow.

11. _____ flows in one direction.

12. _____ is supplied to most residences.

13. Electrons flowing in an electrical circuit are called _____.

14. Current flow is measured in _____.

15. The opposition to electron flow is _____.

16. _____ is the rate at which electrons do work.

17. The measurement for power in an electric circuit is the _____.

18. Electrical utilities bill customers for electrical service by the _____.

19. The _____ is the Btu output of the equipment divided by the power input with a seasonal adjustment.

20. The mathematical relationship among the current, electromotive force, and resistance of an electric circuit is known as _____.

CHAPTER 3 Electric Circuits

Chapter Overview

An understanding of electrical circuitry is important to technicians who install and service heating, cooling, and refrigeration equipment and control systems. Any electrical system used in the industry is composed of various types of circuits designed to do specific tasks in the equipment or control system. The three types of electric circuits used in the industry are series, parallel, and series parallel. The series circuit has only one path for electron flow. The parallel circuit has more than one path for electron flow. The series-parallel circuit is a combination of series and parallel circuits.

The series circuit makes up most of the control circuits used in the industry because if any switch in a series circuit opens, the load in the circuit will stop or start, depending on the position of the switches. For example, if any switch that is connected in series with a load is open, the load will be de-energized; if all the switches in the circuit are closed, the load will operate. An operating control such as a thermostat is connected in series with the load that it controls; a closed thermostat will allow the load to operate while an open thermostat will prevent the load from operating. Most safety and operating controls are connected in series to control loads in electrical circuits.

Parallel circuits are used in the industry and in equipment to insure that proper voltage is supplied to the load. Parallel circuits are designed with more than one path for electron flow. In the electrical circuitry for equipment, the parallel circuit is used to supply the correct voltage to the electrical loads in the system. Parallel circuits are used in the wiring of homes to insure that all receptacles are supplied with 115 volts. If an air-conditioning unit has four major loads, all four loads must be supplied with line voltage. The four loads must be wired in parallel to receive the correct voltage.

The series-parallel circuit is a combination of series and parallel circuits. In the proper combination, series-parallel circuits control and supply loads with the correct voltage. Almost all air-conditioning control circuits are series parallel because most control systems or equipment have more than one load that requires line voltage. For example, if three loads require line voltage they will be connected in parallel, but the operating and safety controls will be connected to each load in series. Most electric circuits used to operate loads safely and properly will be of the series-parallel type.

Series and parallel circuits have different voltage, amperage, and resistance relationships. In a series circuit, the voltage is divided among the electrical loads, which is the reason this circuit is not used to supply power to loads in equipment. The current in the series circuit is equal in all parts of the circuit. The sum of all resistances in series is the total circuit resistance. In a parallel circuit, the voltage in all circuit parts is equal, which is the reason that parallel circuits are used in the load circuitry of most equipment. The sum of the current flow in each part of a parallel circuit is the total current. The reciprocal of the total resistance in a parallel circuit is the sum of the reciprocals of all resistances.

Key Terms

Closed circuit
Control circuit
Electric circuit

Open circuit
Parallel circuit
Power circuit

Series circuit
Series-parallel circuit
Voltage drop

REVIEW TEST

Name: _____   Date: _____   Grade: ___

Answer the following questions.

1. What are the requirements for an electric circuit?

2. What happens to the current flow if a switch in the circuit is opened?

3. What type of current is used to supply power to HVAC (heating, ventilation, and air-conditioning) equipment?

4. Name some electrical components that use direct current in control systems of HVAC equipment.

5. What is a series circuit and how is it used in HVAC equipment?

6. Why are safety controls placed in series with the loads they are protecting?

7. What is a control circuit?

8. Draw a series circuit with a motor being controlled by a thermostat with a high-pressure switch and low-pressure switch as safety devices.

9. Explain the characteristics of the voltage, current, and resistance in a series circuit.

10. What is a parallel circuit and how is it used in HVAC equipment?

11. Draw a parallel circuit with a compressor and condenser fan motor.

12. Explain the characteristics of the voltage, current, and resistance in a parallel circuit.

13. How are series-parallel circuits used in HVAC equipment?

14. Draw a series-parallel circuit with a switch controlling a motor and a thermostat controlling a second motor.

15. If five resistances of equal value are connected in series with 100 volts, what is the voltage drop across each resistance?

16. What is the total resistance of a series circuit with four resistances of 12 ohms, 16 ohms, 20 ohms, and 24 ohms?

17. Why is the current the same in all locations in a series circuit?

18. What is the total resistance of a parallel circuit with resistances of 12 ohms and 24 ohms?

19. What is the total current draw of a parallel circuit with a compressor pulling 22 amps, an indoor fan motor pulling 8 amps, and a condenser fan motor pulling 6 amps?

20. What happens to voltage supplies to loads when they are connected in series and parallel?

LAB 3–1 Series and Parallel Circuits

Name: _____   Date: _____   Grade: ___

Comments:

Objectives: Upon completion of this lab, you should be able to recognize, understand, and draw series, parallel, and series-parallel circuits used in residential HVAC equipment. You will be able to correctly assemble basic series, parallel, and series-parallel circuits and answer questions pertaining to each.

Introduction: You must know how series, parallel, and series-parallel circuits are utilized in control systems and equipment in the industry. You must be able to recognize series, parallel, and series-parallel circuits in modern HVAC wiring diagrams. You must be able to make the correct connections of electrical components when control circuits are being installed or serviced.

Text Reference: Chapter 2

Tools and Materials: The following materials and equipment will be needed to complete this lab exercise.
Screwdriver	5-ampere plug fuse
Wire cutting pliers	Miscellaneous wood screws
Crimpers	Wire and electrical connection components
Cleat receptacles	Low-voltage power cord
Light bulbs	One 12″ × 12″ piece of plywood
Switches	Terminal board

Safety Precautions: In this lab exercise, you will be working with 115-volt circuits. It is very important that you make wiring connections in a neat and orderly fashion to prevent the wires and electrical connections from accidentally touching each other. At no time should you try to connect any component into a circuit with the electrical power being supplied to it; make certain that the electrical power is disconnected. When operating circuits, use caution when opening and closing switches in circuits with power being supplied to them; do not touch any electrical connection when electrical power is supplied to the circuit.

LABORATORY SEQUENCE (mark each box upon completion of task)

A. Series Circuits

☐   1. In the center of the plywood, mount two cleat receptacles side by side. They should be approximately 6 inches apart. Mount the terminal board 4 inches above the cleat receptacles. The plywood board should resemble Figure 3.1.

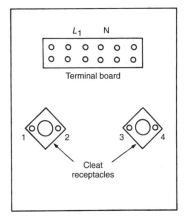

Figure 3.1 Component layout of a basic electrical circuit board.

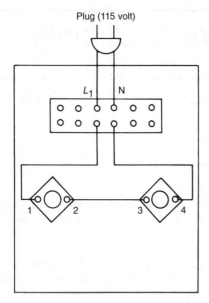

Figure 3.2 Basic electrical circuit board with a series circuit.

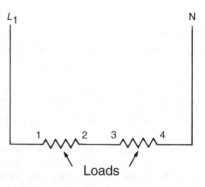

Figure 3.3 Schematic diagram of a basic electrical circuit board with a series circuit.

☐   2. Wire the board as shown in Figure 3.2. The two cleat receptacles form a series circuit, shown schematically in Figure 3.3.

☐   3. Connect the test cord with the 115-volt plug to terminals L_1 and N of the terminal board.

☐   4. Have the instructor check your series circuit.

☐   5. Install two 60-watt light bulbs in the cleat receptacles.

☐   6. Plug the circuit into a 115-volt receptacle. Do the light bulbs burn correctly?

       What is the voltage being supplied to each light bulb?   Why?

☐   7. Unplug the circuit.

☐   8. Install one 60-watt light bulb and one 100-watt light bulb in the cleat receptacles.

☐   9. Plug the circuit into a 115-volt receptacle.

☐   10. Unplug the circuit. Do the light bulbs burn correctly?

       Which light bulb burns the brightest?   Why?

☐   11. Install one 60-watt light bulb and one plug fuse in the cleat receptacles.

☐   12. Unplug the circuit. Why does the 60-watt light bulb now burn correctly?

       Explain how the circuit operates.

B. Parallel Circuits

☐   1. Wire the board as shown in Figure 3.4. The two cleat receptacles are wired in parallel, as shown schematically in Figure 3.5.

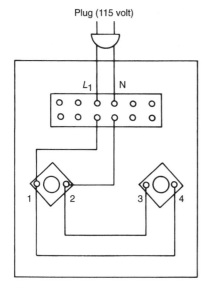

Plug (115 volt)

L_1 N

1   2   3   4

Figure 3.4 Basic electrical circuit board with a parallel circuit.

L_1 N

1   2

3   4

Loads

Figure 3.5 Schematic diagram of a basic electrical circuit board with a parallel circuit.

2. Have your instructor check your parallel circuit.

3. Install two 60-watt light bulbs in the cleat receptacles.

4. Plug the circuit into a 115-volt receptacle. Do the light bulbs burn correctly?   Why?

What voltage is being supplied to each light bulb?

What would the results be if we changed the wattage of the light bulbs?

5. Unplug the circuit.

C. Series-Parallel Circuits

1. The series-parallel circuit is a combination of the series and parallel circuits. The electrical circuitry of most air-conditioning equipment is composed of series-parallel circuits. An example of a series-parallel circuit is shown in Figure 3.6.

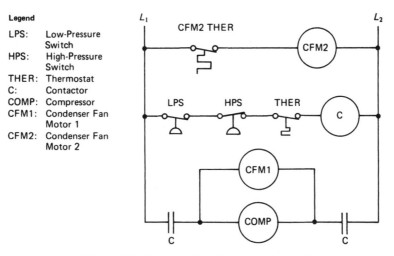

Legend

LPS: Low-Pressure Switch
HPS: High-Pressure Switch
THER: Thermostat
C: Contactor
COMP: Compressor
CFM1:   Condenser Fan Motor 1
CFM2:   Condenser Fan Motor 2

Figure 3.6 Example of series-parallel circuit.

2. Draw and wire the schematic diagrams of the following series-parallel circuits.

☐   a.   Circuit A: Connect two light bulbs in parallel and control each bulb by a switch in series with each light bulb.

       b.   Draw the circuit.

☐   c.   Have your instructor check your drawing.

☐   d.   Wire and operate the circuit.

           What happens when the switches are closed?

           What happens when the switches are opened?

☐   e.   Circuit B: Connect two light bulbs in parallel and control both light bulbs with one switch.

☐   f.   Draw the circuit.

☐   g.   Have your instructor check your drawing.

☐   h.   Wire and operate the circuit.

            What happens when the switch is closed?

            What voltage is being supplied to the light bulbs?

☐   i.   Circuit C: Connect two light bulbs in series with one switch and connect another switch in parallel with one light bulb.

☐   j.   Draw the circuit.

☐   k.   Have your instructor check your drawing.

☐   l.   Wire and operate the circuit.

            Explain the operation of the switch wired in parallel to the light bulb.

☐   m.   Disconnect all wiring from components and remove the components from the plywood and return to their proper location(s).

SUMMARY STATEMENT: Describe the application of series, parallel, and series-parallel circuits in HVAC control systems and equipment.

Questions

1. Two 115-volt light bulbs connected in series with 115 volts would _____.
   a. burn out immediately
   b. burn correctly
   c. burn dimly
   d. do none of the above

2. Two 115-volt light bulbs connected in parallel with 115 volts would _____.
   a. burn out immediately
   b. burn correctly
   c. burn dimly
   d. do none of the above

3. A thermostat, high-pressure switch, and low-pressure switch are connected in series with a small compressor. If the low-pressure switch opened, the compressor would _____.
   a. continue to run
   b. stop immediately
   c. stop if all three safety switches opened
   d. do none of the above

4. Four light bulbs of the same wattage are connected in parallel with 115 volts. The voltage at each bulb is _____.
   a. 115   c. 27.5
   b. 55   d. 0

5. Four light bulbs are connected in series with 440 volts. The voltage at each bulb is _____.
   a. 440   c. 110
   b. 230   d. 55

Answer the following questions using Figure 3.7.

6. The compressor and condenser fan are connected in _____.
   a. parallel   b. series

7. The high-pressure switch and low-pressure switch are connected in _____.
   a. parallel   b. series

8. The indoor fan relay contacts are connected in _____ with the indoor fan motor.
   a. series   b. parallel

9. The indoor fan relay coil and the contactor coil are connected in _____.
   a. series   b. parallel

10. The compressor and compressor running capacitor are connected in _____.
    a. series   b. parallel

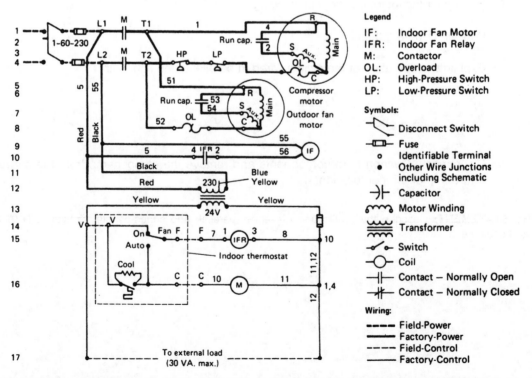

Figure 3.7 Complete schematic diagram for a small packaged unit. *(Courtesy of Westinghouse Electric Corp., Central Residential Air Conditioning Division)*

CHAPTER 4 | Electric Meters

Chapter Overview

The installation and service technician in the heating, cooling, and refrigeration industry will often be required to use electrical meters to perform the tasks that are required when installing, testing, and servicing equipment. When installing equipment, the installation technician will often need to check the voltage supplied to the equipment and the current draw of the equipment, which helps determine if the equipment is operating properly. The service technician will use electrical meters while troubleshooting systems. As important as electricity is to the HVAC industry, it is not hard to see that it is of utmost importance for service and installation technicians to be able to determine the condition of electrical components and circuits in equipment and control systems. The HVAC technician commonly uses ammeters, voltmeters, and ohmmeters or a combination of any of the three.

The ammeter is used to measure the current flow in an electric circuit. Two types of ammeters, clamp-on and inline, are used in the industry today. The inline ammeter is seldom used because it must be put in series with the load or circuit to read the amperage. Many volt-ohm-milli-ammeters have a milli-amp function that is used to test the ignition circuits of gas furnaces. The clamp-on ammeter is the most common because it is easy to use. To use the clamp-on ammeter, it is only necessary to clamp the jaws of the meter around the conductor feeding the circuit or component and then read the amperage.

The voltmeter is used to measure the voltage of an electric circuit. The voltmeter will have to be connected in parallel with the circuit to determine the voltage supplied to the circuit. Voltage measurements can be used to determine the source voltage, voltage drop, and voltage imbalance. The first step in troubleshooting an air-conditioning system is to determine if voltage is available to the equipment or system.

The ohmmeter is used to measure the resistance of a circuit or device in ohms. The ohmmeter must be used with the circuit power off to prevent damage to the meter. Most ohmmeters have a power source built into the meter. The ohmmeter is used to determine the condition of electrical devices used in air conditioning, such as motors, heaters, contactor coils, solenoid valves, and other components. A continuity check is another use for an ohmmeter. This test determines if a complete path is available for current flow.

Electrical meters come in a variety of sizes, shapes, and configurations. Many meters combine several functions into one meter. The accuracy and decrease in cost of digital meters have made them more popular in the industry. Digital meters can be used without any interpolation of the reading, which gives them an advantage over analog meters, where interpolation must be made with reference to the needle on the scale. Technicians should choose the meter that best suits their needs in the industry. With the proper care, meters will provide many years of service.

Key Terms

Ammeter
Analog meters
Clamp-on ammeter
Continuity

Digital meters
Magnetic field
Measurable resistance
Ohmmeter

Open
Short
Voltmeter

REVIEW TEST

Name: _____ Date: _____ Grade: ___

Complete the following multiple choice questions by circling the correct answer.

1. Most analog electric meters make use of the _____ to create needle movement.
   a. volt wave
   b. magnetic field
   c. sine wave
   d. magnetic flux

2. Which of the following is produced around a conductor when current is flowing through the conductor?
   a. voltage
   b. resistance
   c. amperage
   d. magnetism

3. Why are digital meters becoming more popular in the industry?
   a. reduction in cost
   b. easier to read
   c. accuracy
   d. all of the above

4. A reading on a digital meter is 120.8; this meter utilizes a _____ digit reading.
   a. 3
   b. 3½
   c. 4
   d. 4½

5. The inline ammeter must be connected in _____ to read the current draw of the motor.
   a. series
   b. parallel

6. What electrical meter is commonly used to determine the current draw of an electric motor?
   a. inline ammeter
   b. clamp-on ammeter
   c. volt-ohm-milliammeter
   d. none of the above

7. What are the results of clamping a clamp-on ammeter around more than one conductor?
   a. the reading must be divided by the number of conductors
   b. the reading must be multiplied by the number of conductors
   c. a zero reading
   d. none of the above

8. The conductor is wrapped around the jaws of the clamp-on ammeter five times. If the reading is 35, what is the total current draw of the load?
   a. 7 amps   c. 35 amps
   b. 175 amps   d. 0 amps

9. When a service technician desires to check an unknown voltage; the voltmeter should be set on the _____ scale.
   a. highest
   b. lowest

10. Air-conditioning equipment is designed to operate at a voltage of plus or minus _____ percent.
   a. 2   c. 10
   b. 5   d. 20

11. The voltmeter is connected to the circuit in _____.
    a. series   c. either a or b
    b. parallel   d. none of the above

12. What precaution should be taken when checking the resistance of a circuit with an ohmmeter?
    a. disconnect the circuit from the power source
    b. set the ohmmeter on the highest scale
    c. make sure the fuses are in good condition
    d. none of the above

13. What would be the resistance of a circuit that is open?
    a. 0 ohm   d. infinite ohms
    b. 3 ohms   e. both b and c
    c. 6 ohms

14. What would be the resistance of a motor winding that is shorted?
    a. 0 ohm   d. infinite ohms
    b. 3 ohms   e. both b and c
    c. 6 ohms

15. What would be the resistance of a circuit with a measurable resistance?
    a. 0 ohm   d. infinite ohms
    b. 3 ohms   e. both b and c
    c. 6 ohms

16. What word is used to refer to a complete circuit that allows current to flow?
    a. open   c. continuity
    b. closed   d. direct

17. The digital meter makes use of _____ to measure and display the electrical characteristics of a circuit.
    a. Watt's law   c. Ohm's law
    b. Kirchoff's law   d. none of the above

18. Which of the following steps is the most logical in troubleshooting a unit that will not operate?
    a. check the voltage being supplied to the unit
    b. check the amp draw of the compressor
    c. check the resistance of the compressor
    d. none of the above

19. What would be the results of checking voltage with an ohmmeter that has no safety devices?
    a. nothing
    b. a damaged meter
    c. destroyed meter leads
    d. none of the above

20. If the technician needs to check the resistance of an electrical component in a circuit, what must be done to the component?
    a. check with meter selector switch on $R \times 10,000$ scale
    b. check with meter selector switch on $R \times 1000$ scale
    c. check with meter selector switch on $R \times 1$ scale
    d. isolate the component being checked

LAB 4–1 Reading Electric Meters

Name: _____	Date: _____	Grade: ___

Comments:

Objectives: Upon completion of this lab, you should be able to read the voltage, current, and resistance of electrical circuits and components with analog and digital meters.

Introduction: In all phases of the HVAC industry, technicians will be required to measure the characteristics of an electrical circuit or component in order to troubleshoot or check the unit upon initial starting. There are various types of electric meters and accessories that are used by technicians to accomplish this task. The analog meter has a scale and a needle that determines the reading while a digital meter digitally reads the electrical characteristics being measured. Meters are available in many different designs.

Text Reference: Chapter 3

Tools and Materials: The following materials and equipment will be needed to complete the lab exercise.

Analog volt-ohmmeter	Potential relays (Groups 1–6 coil)
Digital volt-ohmmeter	Voltage relays (230/115/24-volt coils)
Analog clamp-on ammeter	Other miscellaneous controls
Digital clamp-on ammeter	Operating motors and systems

Safety Precautions: Use caution when checking characteristics of an electrical circuit supplied with voltage. Make sure no current-carrying conductors are touching metal surfaces except the grounding conductor. Make sure body parts do not come in contact with live electrical circuits. Keep hands and materials away from moving parts. Do not check the resistance of a circuit with power applied.

LABORATORY SEQUENCE (mark each box upon completion of task)

Before beginning, you should know the type of electrical reading being made and adjust the meter to the correct range and scale.

A. Reading Voltage

Air-conditioning loads and control circuits require a specific voltage for proper operation. Most control systems need 24 volts and the major load, the compressor, needs 115, 208, 230, or 460 volts. In this section, you will read control system voltages used by important system loads. Using analog and digital volt-ohmmeters, record the voltages applied to the following components.

☐   1. Check the following with instructor supervision:

Wall receptacle _____ volts

Compressor _____ volts

Evaporator fan motor _____ volts

Condenser fan motor _____ volts

Transformer output _____ volts

☐   2. Check the following without instructor supervision:

        Electric heater _____ volts

        Wall receptacle _____ volts

        Compressor #1 _____ volts

        Compressor #2 _____ volts

        Evaporator fan motor _____ volts

        Electric motor _____ volts

        Transformer output _____ volts

        Transformer input _____ volts

        Light bulb _____ volts

        Furnace _____ volts

☐   3. Have your instructor check the voltage readings.

B. Reading Amperage

The current flow in an electrical circuit is measured with an ammeter. The current flow used by an electrical load will often give an indication of the condition of the component. Using analog and digital clamp-on ammeters, record the current draw of the following components.

☐   1. Check the following with instructor supervision:

        Compressor _____ amps

        Refrigerator _____ amps

        Electric heater _____ amps

        Evaporator fan motor _____ amps

        Condenser fan motor _____ amps

☐   2. Check the following without instructor supervision:

        Compressor #1 _____ amps

        Compressor #2 _____ amps

        Electric heater _____ amps

        Evaporator fan motor _____ amps

        Electric motor _____ amps

        Condenser fan motor _____ amps

        Electrical panel _____ amps

        Transformer output _____ amps

        Pump motor _____ amps

        Large compressor _____ amps

☐   3. Have your instructor check the amperage readings.

C. Reading Resistance

Resistance is the opposition to electron flow in an electrical circuit. All loads will have a specific resistance. The electrical condition of a load can be determined by reading the resistance of a load. When checking the resistance of a component, make certain the power is disconnected from the power source. Using analog and digital volt-ohmmeters, record the resistance of the following components.

☐   1. Check the following with instructor supervision:

Electric heater _____ ohms

Electric motor _____ ohms

Relay coil (24 volts) _____ ohms

Evaporator fan motor _____ ohms

Resistor _____ ohms

☐   2. Check the following without instructor supervision:

Compressor _____ ohms

Evaporator fan motor _____ ohms

Condenser fan motor _____ ohms

Transformer _____ ohms

Relay coil (115 volts) _____ ohms

Relay coil (230 volts) _____ ohms

Potential relay coil #1 _____ ohms

Potential relay coil #2 _____ ohms

Potential relay coil #3 _____ ohms

Potential relay coil #4 _____ ohms

☐   3. Have your instructor check the resistance readings.

D. Circuit Characteristics

☐   1. Measure and record the following circuit characteristics:

Compressor _____ volts _____ amps _____ ohms

Electric heater _____ volts _____ amps _____ ohms

Electric motor _____ volts _____ amps _____ ohms

E fan motor _____ volts _____ amps _____ ohms

C fan motor _____ volts _____ amps _____ ohms

☐   2. Have your instructor check your readings.

SUMMARY STATEMENT: Explain the types of electric meters that a service technician would need to adequately troubleshoot an inoperative air-conditioning system.

Questions

1. At what point on the scale does an electric meter read most accurately?

2. If you are checking the voltage available to an air-conditioning unit but have no idea what it could be, what scale should you use?

3. If you need to check the current draw of a very small load (approximately .5 amp), how could you accurately measure the amperage of the load?

4. If an ohmmeter has the following resistance scales, $R \times 1$, $R \times 10$, $R \times 100$, and $R \times 1000$, which scale would a technician use to measure a resistance of 26,000 ohms?

5. What methods are used to protect the electrical circuitry of a digital meter?

6. How does a clamp-on ammeter operate?

7. Give the advantages of a digital electric meter.

8. What is the meaning of continuity in reference to an electrical circuit?

9. What is the difference between a short and an open circuit?

10. If you were going to purchase an electric meter to be used in the HVAC industry, which type of meter would you purchase and why?

CHAPTER 5 — Components, Symbols, and Circuitry of Air-Conditioning Diagrams

Chapter Overview

Approximately 85% of the problems in refrigeration, heating, and air-conditioning electrical systems are electrical. In order to effectively troubleshoot the electrical devices and control systems of modern HVAC equipment, the technician must know how to read and interpret electrical wiring diagrams. Electrical diagrams contain a wealth of information about the electrical installation and operation of the equipment. The installation technician depends on the wiring diagram for assistance in making the proper electrical connections. The service technician uses schematic diagrams as a guide in troubleshooting the electrical control system of refrigeration, heating, and air-conditioning systems and equipment. The schematic diagram is broken down into a circuit-by-circuit arrangement that allows the technician to easily identify the electrical circuit that is causing the problems. Once the inoperative circuit is located, the technician can determine which electrical component is at fault and must be replaced. The efficient use of wiring diagrams decreases the amount of time required in the troubleshooting process.

It would be impossible for electrical components to be represented in a wiring diagram by photographs because of their size and the complexity that would result in showing the wiring connections made to each component. Thus, symbols are used in wiring diagrams to represent the electrical components, such as thermostats, pressure switches, overloads, transformers, contactors, relays, and motors. Technicians must be able to identify most symbols and know where to look up the remainder. Symbols are not standard across the industry; therefore, the technician must have some means of knowing what a symbol represents. The legend of the wiring diagram is a listing of the components along with their symbols.

Electrical devices generally can be divided into two basic types, loads and switches. Loads are electrical devices that consume electrical energy to do useful work. Common loads in control systems are motors, solenoids, resistance heaters, and other current-consuming devices. Switches are electrical devices that open and close due to changes in the medium that controls the switch, such as temperature, pressure, or a solenoid coil. In an electrical diagram, loads are controlled by switches to maintain a certain condition. There are various devices that don't neatly fit into either the switch or load categories, but in most cases they are classified by this method.

Key Terms

Contactor	Magnetic overload	Push-button
De-energized	Magnetic starter	Relay
Disconnect switch	Motor	Schematic diagram
Energized	Normally	Signal light
Factual diagram	Normally closed	Solenoid
Fuse	Normally open	Switch
Heat pump	Pictorial diagram	Thermal overload
Heater	Pilot duty	Thermostat
Installation diagram	Pole	Throw
Load	Pressure switch	Transformer

REVIEW TEST

Name: _____   Date: _____   Grade: ____

Match the following terms and definitions.

_____ 1. Legend

    a. A device that decreases the input voltage

_____ 2. Motor

    b. A device that opens and closes a set of contacts when its coil is energized; usually carries less than 20 amperes

_____ 3. Thermostat

    c. Cross-references the components and their letter designations

_____ 4. Fuse

    d. Simplest type of overload

_____ 5. Pilot duty

    e. Heats a conditioned space by reversing the refrigeration cycle

_____ 6. Load

    f. A device such as a motor, heater, signal light, or solenoid

_____ 7. Switch

    g. A device that breaks a circuit when a high current draw exists

_____ 8. Contactor

    h. An electrical device that is used to rotate a compressor

_____ 9. Relay

    i. To supply energy

_____ 10. Magnetic starter

    j. An electrical device that opens and closes to control a load

_____ 11. Normally closed

    k. One set of electrical contacts

_____ 12. Throw

    l. Similar to a contactor except with a means of overload protection

_____ 13. Pole

    m. Also called a line or label diagram

_____ 14. Transformer

    n. The position of an electrical device that is open when the device is de-energized

_____ 15. Overload

    o. A device that responds to a temperature change by opening or closing a set of contacts

_____ 16. Schematic diagram

    p. Number of positions of a set of movable contacts

_____ 17. Pictorial diagram

    q. A wiring diagram that is laid out in a circuit-by-circuit arrangement

_____ 18. Energize

    r. The position of an electrical device that is closed when the device is de-energized

_____ 19. Heat pump

    s. A device that opens and closes a set of contacts; capable of carrying loads in excess of 20 amperes

_____ 20. Normally open

    t. A device that breaks the control circuit during a motor overload

Name: _____	Date: _____	Grade: ___

Comments:

Objectives: Upon completion of this lab, you should be able to identify and draw commonly used electrical symbols used in schematic diagrams and be able to locate electrical components from the symbols and wiring diagrams.

Introduction: You must be able to recognize common electrical symbols in order to troubleshoot electrical systems. Although most symbols used in the industry are standard, always check the legend of the wiring diagram for the names of the symbols being used; you will find some differences.

Text References: Paragraphs 5.1 through 5.6

Tools and Materials: The following materials and equipment will be needed to complete this lab exercise.
Selection of common electrical components
Packaged air conditioner and wiring diagram
Heat pump and wiring diagram

Safety Precautions: Make certain that the electrical source is disconnected when identifying electrical components on equipment. Make sure body parts do not come in contact with live electrical conductors. Keep hands and materials away from moving parts.

LABORATORY SEQUENCE (mark each box upon completion of task)

☐ **A. Symbol Identification**

On the figure shown below, write the name of the device next to its associated symbol.

☐ **B. Drawing Symbols**

Draw the symbols for the components listed below.

1. Fuse

11. Magnetic overload

2. Red signal light

12. Transformer

3. Compressor motor

13. Capacitor

4. Three-speed indoor fan motor

14. Low-pressure switch (closes on rise in pressure)

5. Reversing valve solenoid

15. High-pressure switch (opens on rise in pressure)

6. Crankcase heater

16. Thermal line break overload

7. Heating thermostat

17. Hot gas solenoid

8. Cooling thermostat

18. Supplementary heater

9. Three-phase compressor motor

19. Three-pole contactor

10. Relay with one NO contact and one NC contact

20. Magnetic starter

C. Drawing Symbols for Electrical Components

☐ 1. Obtain ten different electrical components.

☐ 2. Draw the symbols for each component.

    a.   f.

    b.   g.

    c.   h.

    d.   i.

    e.   j.

☐ 3. Have your instructor check the symbols.

D. Locating Components from Symbols

☐ 1. Your instructor will assign an air-cooled packaged unit on which to identify the following components:

    a. Condenser fan motor   e. Thermostat
    b. Evaporator fan motor   f. Capacitor
    c. Compressor motor   g. Indoor fan relay
    d. Transformer   h. Contactor

☐ 2. Locate the following components on the air-to-air heat pump assigned by your instructor:

    a. Reversing valve solenoid   f. Contactor
    b. Electrical resistance heater   g. Thermal overload for electric heaters
    c. Low-pressure switch   h. Thermostat
    d. Crankcase heater   i. Outdoor fan motor
    e Defrost relay   j. Defrost thermostat

MAINTENANCE OF WORK STATION AND TOOLS: Clean and return all tools to their proper location(s). Return all electrical components to their proper location(s). Replace all covers on equipments used in this exercise.

SUMMARY STATEMENT: Describe the use and purpose of letter designations on electrical symbols.

Questions

1. Draw a relay with two NO contacts and one NC contact in the de-energized and energized positions.

2. Draw the symbol for and explain the action of a single-pole–double-throw switch.

3. What is the difference between a magnetic starter and a contactor?

4. Draw the symbols for the following motors. (Use letter designations.)

   a. compressor
   b. indoor fan motor
   c. timer motor

   d. condenser fan motor
   e. condensate pump motor

5. Name at least five electrical loads that are used in the HVAC industry.

6. What is the purpose of the disconnect switch?

7. Draw the symbols for heating and cooling thermostats and explain the difference.

8. How can you differentiate between a thermostat and a pressure switch without a legend?

9. Draw the symbol for a thermal and magnetic overload and explain their controlling elements.

10. Draw the symbol for a 208/230-volt transformer.

CHAPTER 6 Reading Schematic Wiring Diagrams

Chapter Overview

Many types of electrical diagrams are used in the refrigeration, heating, and air-conditioning industry. While all types of electrical diagrams are important, the schematic diagram is the most important to the service technician. The schematic diagram is broken down into a circuit-by-circuit arrangement. The schematic diagram is sometimes referred to as a ladder diagram. In a schematic diagram, the power source is represented by two vertical lines, with the circuits of the equipment or control system connected between them. If a service technician is able to break down a control system into a circuit-by-circuit arrangement, diagnosis of system problems is made much easier. Technicians must understand the operation of equipment they are assigned to troubleshoot. A proficient technician should be able to read and interpret schematic diagrams in order to (1) understand how a piece of equipment operates, (2) know what the unit should be doing when it is in a specific mode of operation, (3) be able to make installations connections, and (4) be able to troubleshoot and repair refrigeration, heating, and air-conditioning systems. Diagrams can usually be found attached to one of the control panels of the equipment. The schematic diagram tells the technician how, when, and why a system works as it does.

A schematic diagram uses symbols to represent electrical components in the control system. The schematic diagram shows the control system in a circuit-by-circuit arrangement with electrical circuits connected between the power source (represented by two vertical lines). A schematic diagram is made up of series and parallel circuits. Each circuit that is connected between the vertical lines or power supply is in parallel. These circuits would supply electrical energy to pass through switches controlling the load in the circuit. The circuit-by-circuit arrangement of schematic diagrams has important benefits because technicians can be overwhelmed when looking at the complexity of an entire diagram. If the technician can isolate one circuit, the diagnosis of problems becomes easier. Isolating a circuit that is causing problems is much easier than trying to deal with the entire

diagram. A technician can never lose sight of the fact that the control circuit operates as a unit, but troubleshooting can be a lot easier if technicians focus on the circuit that is not operating correctly. Technicians must be able to read and interpret schematic diagrams so they can perform tasks that will be assigned to them in the industry.

Control systems sometimes use a transformer in the circuitry to reduce the voltage to a more usable voltage for the control system. Residential control systems and some light commercial systems are usually 24 volts. Large commercial and industrial systems are usually 110 volts. Low-voltage control systems are generally more accurate and can perform more functions as compared to the line voltage control systems. Many schematic diagrams will utilize transformers to reduce this voltage for the control system.

Some schematic diagrams are easy to understand, while others are more difficult. In Chapter 6, schematic diagrams from simple to complex have been covered in detail. The following six control systems are covered in the basic schematic part of the chapter: (1) dehumidifier, (2) simple window air conditioner, (3) walk-in cooler, (4) commercial freezer, (5) gas furnace with standing pilot, and (6) package air conditioner. The following six control systems are covered in the advanced schematic part of the chapter: (1) light commercial air-conditioning control system with a control relay, (2) light commercial air-conditioning control system with a lockout relay, (3) two-stage heating, two-stage cooling control system, (4) heat pump with defrost timer, (5) heat pump with defrost board, and (6) commercial refrigeration system using a pump-down control system.

In Chapter 6, the schematics of several different applications were covered in detail from the simple to the complex. Most of the time, if the technician can isolate the circuit that is creating problems, the task of diagnosing the problem will be much easier. No matter what type of control system is used in a heating, air-conditioning, or refrigeration system, it is essential that technicians be able to read schematic diagrams so they can efficiently troubleshoot the system.

Key Terms

Balance point
Combustion chamber
Control relay
Defrost cycle
Dehumidifier
Gas furnace

Heat pump
Light commercial air-
 conditioning system
Limit switch
Line voltage control system
Lockout relay

Low-voltage control system
Multistage thermostat
Pump-down control system
Reversing valve
Set point
Short cycling

REVIEW TEST

Name: _____   Date: _____   Grade: _____

Answer the following questions.

1. Why is it important for a service technician to be able to read and interpret schematic diagrams?

2. Briefly explain the design of a schematic wiring diagram.

3. What is the difference between a line voltage and low-voltage control system?

4. How are parallel circuits utilized in a schematic diagram?

5. How are series circuits utilized in a schematic diagram?

6. What is a dehumidifier?

7. What is the operating sequence of the window air conditioner shown in Figure 6.1?

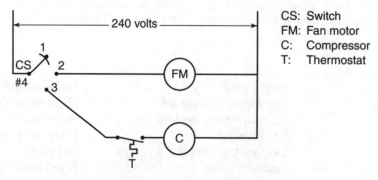

CS: Switch
FM: Fan motor
C: Compressor
T: Thermostat

Figure 6.1 Schematic diagram of a simple window unit.

Electricity for Refrigeration, Heating, and Air Conditioning Lab Manual, **Eighth Edition**

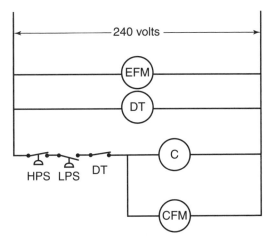

EFM: Evaporator fan motor
C: Compressor
LPS: Low-pressure switch
HPS: High-pressure switch
CFM: Condenser fan motor
DT: Defrost timer

Figure 6.2 Schematic diagram of a walk-in cooler.

Answer questions 8–11 based on Figure 6.2.

8. The EFM, DT, C, and CFM are connected in _____.
   a. series
   b. parallel

9. The HPS, LPS, DT, and C are connected in _____.
   a. series
   b. parallel

10. The EFM and DT operate _____.
   a. when power is supplied to the system
   b. when the LPS is open
   c. only when LPS is closed
   d. none of the above

11. What is the purpose of the defrost timer?

Answer questions 12–15 based on Figure 6.3.

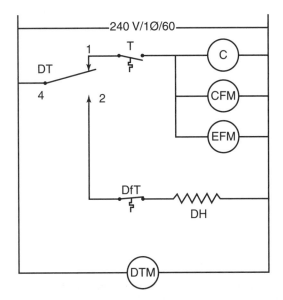

DT: Defrost timer
DTM: Defrost timer motor
T: Thermostat
C: Compressor
CFM: Condenser fan motor
EFM: Evaporator fan motor
DfT: Defrost thermostat
DH: Defrost heater

Figure 6.3 Schematic diagram of a commercial freezer.

12. The C, CFM, and EFM are connected in _____.
    a. series
    b. parallel

13. The DT and DfT are connected in _____ with the DH.
    a. series
    b. parallel

14. With the DT in the position shown, which of the following components will operate?
    a. C
    b. CFM
    c. EFM
    d. all of the above

15. With the defrost timer contacts closed from 4 to 2, what are the results if DfT opens?
    a. CFM will be energized
    b. DH will be de-energized
    c. nothing
    d. T will open

Answer questions 16–18 based on Figure 6.4.

16. What control operates the FM?

17. What is the purpose of the LS?

18. The control system is _____ voltage.

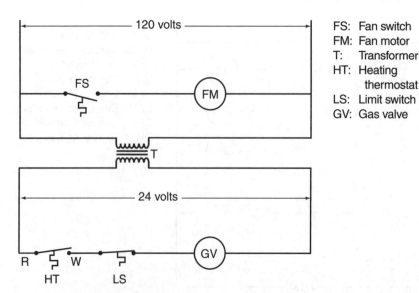

Figure 6.4 Schematic diagram of a gas furnace with a standing pilot.

FS: Fan switch
FM: Fan motor
T: Transformer
HT: Heating thermostat
LS: Limit switch
GV: Gas valve

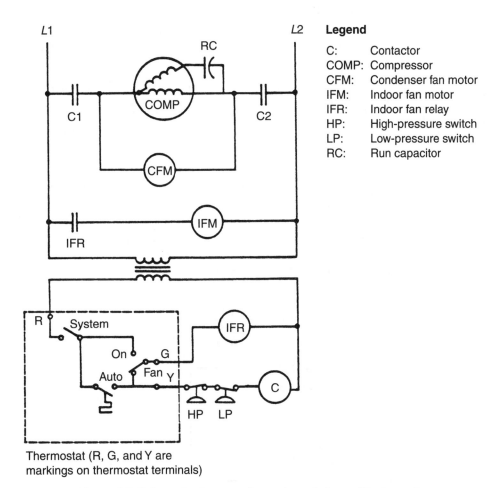

L1   L2

RC

COMP

C1   CFM   C2

IFR   IFM

Thermostat (R, G, and Y are
markings on thermostat terminals)

R
System
On   G
Auto Fan Y
IFR
C
HP LP

Legend

C: Contactor
COMP: Compressor
CFM: Condenser fan motor
IFM: Indoor fan motor
IFR: Indoor fan relay
HP: High-pressure switch
LP: Low-pressure switch
RC: Run capacitor

Figure 6.5 Schematic diagram of a packaged air-conditioning unit.

Answer questions 19–22 based on Figure 6.7.

19. What is the purpose of the control relay and why is it used?

20. If the HPS, LPS, and CIT are closed, what would be the results if CR contacts close?

21. Explain the operation of the CH.

22. When will the CFM operate?

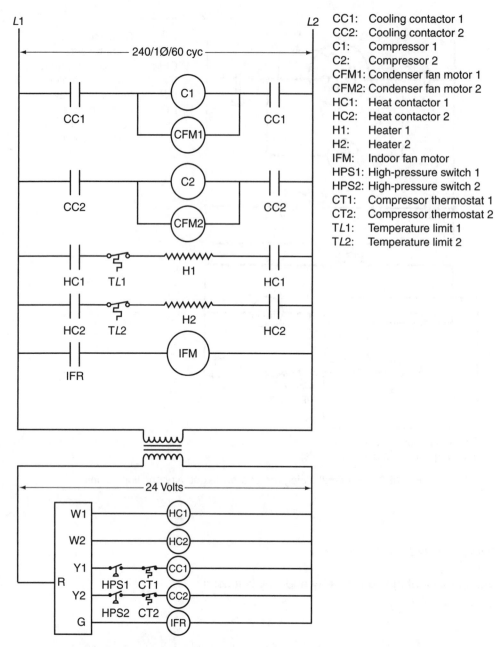

CC1:	Cooling contactor 1
CC2:	Cooling contactor 2
C1:	Compressor 1
C2:	Compressor 2
CFM1:	Condenser fan motor 1
CFM2:	Condenser fan motor 2
HC1:	Heat contactor 1
HC2:	Heat contactor 2
H1:	Heater 1
H2:	Heater 2
IFM:	Indoor fan motor
HPS1:	High-pressure switch 1
HPS2:	High-pressure switch 2
CT1:	Compressor thermostat 1
CT2:	Compressor thermostat 2
TL1:	Temperature limit 1
TL2:	Temperature limit 2

Figure 6.6 Schematic diagram of two-stage cooling, two-stage heating control system.

Answer questions 23–25 based on Figure 6.6.

23. The thermostat used in the control system is a (an) _____.
   a. one-stage cooling, one-stage heating
   b. two-stage cooling, one-stage heating
   c. one-stage cooling, two-stage heating
   d. two-stage heating, two-stage cooling

24. HC2 controls _____.
   a. IFM
   b. H1
   c. H2
   d. all of the above

25. CC1 and CC2 close _____ when operating properly.
   a. at the same time
   b. at different times

LAB 6–1 Reading Basic Schematic Diagrams

Name: _____	Date: _____	Grade: _____

Comments:

Objectives: Upon completion of this lab, you should be able to interpret the operation of the equipment assigned by the instructor and write an operational sequence.

Introduction: A schematic diagram shows a systematic layout of a control system. It can tell you how, when, and why a system works as it does. Reading the diagram is easy if you take one circuit at a time instead of trying to follow the entire diagram at once. All schematic diagrams are broken down into basic circuits, and each circuit usually contains one load.

Text Reference: Paragraphs 6.1 and 6.2

Tools and Materials: Selection of wiring diagrams

LABORATORY SEQUENCE (mark each box upon completion of task)

☐ **A. Simple Window Air Conditioner**

Answer the following questions using Figure 6.1.

1. The switch is in _____ to the fan motor and compressor.
   a. series
   b. parallel

2. What component will be energized with the switch making connections between #4 and #2?
   a. fan motor
   b. compressor
   c. thermostat
   d. none

3. Which of the following components must be closed for the compressor to be energized?
   a. switch (#4 to #2) and thermostat
   b. switch (#4 to #3) and thermostat
   c. switch (#4 to #3)
   d. thermostat

4. The fan motor and compressor are connected in _____.
   a. parallel
   b. series

5. Write an operational sequence for the window unit in Figure 6.1.

NOTE: An operational sequence is a brief description of the how a system operates electrically.

☐ **B. Walk-in Cooler**

Answer the following questions using Figure 6.2.

1. The operating control of the walk-in cooler is
   a. the high-pressure switch
   b. the low-pressure switch
   c. the defrost timer contacts
   d. none of the above

2. Which of the following components operate continuously?
   a. EFM and DT motor
   b. EFM and C
   c. C and CFM
   d. DT and CFM

3. Which of the following controls is used as a safety device?
   a. DT
   b. LPS
   c. HPS
   d. all of the above

4. Which of the following components is not in parallel with the compressor?
   a. EFM
   b. CFM
   c. DT
   d. LPS

5. Write an operational sequence for the walk-in cooler in Figure 6.2.

☐ **C. Commercial Freezer**

Answer the following questions using Figure 6.3.

1. What is the operating control of the freezer?

2. What components of the system are energized and de-energized when the system goes into defrost?

3. What component de-energizes the defrost heater if the evaporator coil of the freezer becomes too warm?

4. Explain the defrost cycle of the freezer.

5. Explain the operating cycle of the freezer.

☐ **D. Gas Furnace with Standing Pilot**

Answer the following questions from Figure 6.4.

1. The control system used in the gas furnace is
   a. low voltage
   b. line voltage

2. The FM operates when
   a. the thermostat closes
   b. the GV is energized
   c. power is supplied to the furnace
   d. the fan switch is heated and closes

3. If the LS opens, the GV is _____.
   a. energized
   b. de-energized

4. The limit switch opens when
   a. the FM overheats
   b. the furnace combustion chamber overheats
   c. the GV overheats
   d. none of the above

5. Write an operational sequence for the furnace in Figure 6.4.

☐ **E. Small Packaged Unit**

Answer the following questions using Figure 6.5.

1. The HP and LP switches are connected in _____ with the compressor motor.
   a. series
   b. parallel

2. The compressor motor is protected by an _____ overload.
   a. internal
   b. external

3. The condenser fan motor is protected by an _____ overload.
   a. internal
   b. external

4. The contactor is controlled by _____.
   a. the disconnect switch
   b. the high-pressure switch
   c. the cooling thermostat
   d. none of the above

5. With the fan switch in the "on" position, the indoor fan will _____.
   a. run when the cooling thermostat is closed only
   b. run when the temperature exceeds 95°F
   c. run continuously
   d. do none of the above

6. The compressor and condenser fan motor start _____.
   a. at different times
   b. at the same time

7. The compressor and condenser fan motor are connected in _____.
   a. series
   b. parallel

8. Write an operational sequence for the packaged unit in Figure 6.5.

Questions

1. A schematic diagram shows the components in _____.
   a. an energized position
   b. a de-energized position

2. A schematic diagram is _____.
   a. an exact picture of a control panel with connecting wire
   b. an installation diagram
   c. a circuit-by-circuit layout of the control panel
   d. none of the above

3. A schematic diagram is most important for _____.
   a. servicing air-conditioning control systems
   b. locating components in the control panel
   c. installing air-conditioning equipment
   d. the homeowner

4. Most loads shown in a schematic diagrams are connected in _____.
   a. series
   b. parallel

5. Most switches shown in schematic diagrams are connected in _____.
   a. series
   b. parallel

6. What is the operating control in the commercial freezer in Figure 6.3?
   a. DT contacts
   b. T
   c. DfT
   d. DH

7. What components are connected in parallel in the commercial freezer in Figure 6.3?
   a. C
   b. CFM
   c. EFM
   d. all of the above

8. What is the switch configuration of the DT in Figure 6.3?
   a. DPDT
   b. DPST
   c. SPST
   d. SPDT

9. The control system of the commercial freezer in Figure 6.3 is a _____ voltage.
   a. line
   b. low

10. Why is it important for the service technician to be able to read schematic diagrams?

Reading Advanced Schematic Diagrams

Name: _____	Date: _____	Grade: ____

Comments:

Objectives: Upon completion of this lab, you should be able to interpret the operation of the equipment assigned by the instructor and write an operational sequence.

Introduction: A schematic diagram shows a systematic layout of a control system. It can tell you how, when, and why a system works as it does. Reading the diagram is easy if you take one circuit at a time instead of trying to follow the entire diagram at once. All schematic diagrams are broken down into basic circuits, and each circuit usually contains one load.

Text References: Paragraphs 6.1 through 6.3

Tools and Materials: Selection of wiring diagrams

LABORATORY SEQUENCE (mark each box upon completion of task)

☐ **A. Light Commercial Packaged Air Conditioner with Control Relay**

Answer the following questions using Figure 6.7.

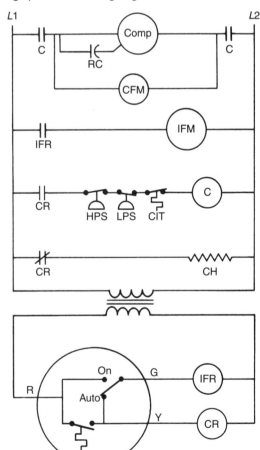

Legend

COMP:	Compressor
C:	Contactor
IFR:	Indoor fan relay
IFM:	Indoor fan motor
CR:	Control relay
HPS:	High-pressure switch
LPS:	Low-pressure switch
CR:	Control relay
CH:	Crankcase heater
TRANS:	Transformer
CIT:	Compressor internal thermostat
CT:	Cool thermostat

Figure 6.7 Schematic diagram of light commercial air-conditioning unit with control relay.

1. The crankcase heater is _____ when the compressor is operating.
   a. energized
   b. de-energized

2. What are the results if the HPS opens while the system is operating?
   a. the condenser fan motor is de-energized
   b. the compressor is de-energized
   c. the crankcase heater is energized
   d. all of the above

3. The IFR controls the _____.
   a. Comp
   b. CFM
   c. IFM
   d. CH

4. If the transformer was faulty, which of the following components would operate properly?
   a. IFM
   b. Comp
   c. CFM
   d. CH

5. Which of the following devices are connected in parallel?
   a. CFM and C contacts
   b. HPS and CIT
   c. CR normally closed contacts and CH
   d. IFM & C

6. Why would a control relay be used in a control system?
   a. a control system with a large number of large 24 volt loads
   b. a 240 volt control system
   c. a residential air conditioner
   d. none of the above

7. The IFM would operate if the CR failed to close.
   a. true
   b. false

8. The CIT would open on a (an) _____ in compressor motor temperature.
   a. increase
   b. decrease

9. Write an operational sequence for the light commercial air conditioner in Figure 6.7.

Answer the following questions using Figure 6.8.

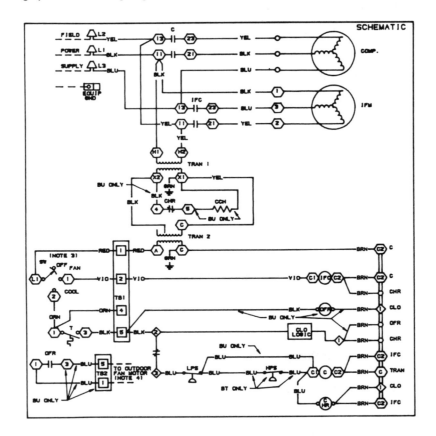

LEGEND

C	— Compressor Contactor
CH or CCH	— Crankcase Heater
CHR	— Crankcase Heater Relay
CLO	— Compressor Lockout
COMP	— Compressor
CR	— Control Relay
DU	— Dummy Terminal
Equip Gnd	— Equipment Ground
HPS	— High Pressure Switch
IFC	— Indoor Fan Contactor
IFM	— Indoor Fan Motor
IP	— Internal Protector
LLS	— Liquid Line Solenoid
LPS	— Low Pressure Switch
OFR	— Outdoor Fan Relay
OL	— Overload
QT	— Quadruple Terminal
RC	— Run Capacitor
S	— Compressor Solenoid
SC	— Start Capacitor
SR	— Start Relay
SW	— Switch
T	— Thermostat
TB	— Terminal Block (Board)
TC	— Thermostat Cooling
Tran	— Transformer

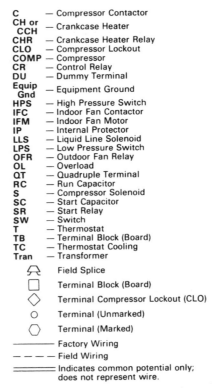

⌒	Field Splice
☐	Terminal Block (Board)
◇	Terminal Compressor Lockout (CLO)
○	Terminal (Unmarked)
⬡	Terminal (Marked)
——	Factory Wiring
— — —	Field Wiring
══	Indicates common potential only; does not represent wire.

Figure 6.8 Schematic diagram of an air-cooled packaged unit with a remote condenser to be used with Part B questions. *(Reproduced courtesy of Carrier Corporation, Syracuse, NY)*

1. True or False.   The compressor and indoor fan motors are single-phase motors.

2. True or False.   The crankcase heater is de-energized when the compressor is operating.

3. True or False.   The HPS closes on a rise in pressure.

4. True or False.   The contacts of the compressor lockout are normally closed.

5. True or False.   The outdoor fan relay is connected in parallel with the contactor coil.

6. True or False.   The "off/fan/cool" switch will allow compressor operation without indoor fan operation.

7. True or False.   The "off/fan/cool" switch controls the operation of the indoor fan motor and compressor.

8. True or False.   The outdoor fan motor is supplied power from L1 and L2 in this schematic.

9. Write an operational sequence for the air-cooled packaged unit in Figure 6.8.

10. Have your instructor check your answers to questions 1–9.

☐ **C. Commercial Freezer with a Pump-Down Control**

Answer the following questions using Figure 6.9.

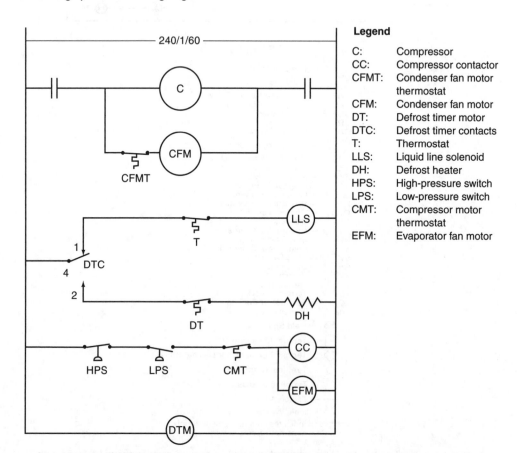

Figure 6.9 Schematic diagram of a commercial freezer with pump-down control.

1. What controls the LLS?
   a. CFMT
   b. T
   c. DT
   d. LPS

2. Which of the following components operate continuously?
   a. CFM
   b. DTM
   c. CC
   d. LLS

3. Which of the following components operate at the same time?
   a. C
   b. LLS
   c. EFM
   d. all of the above

4. What component is energized if the DTC contacts are from 4 to 2 and the DT is closed?
   a. DH
   b. LLS
   c. CC
   d. C

5. If the compressor is operating and the condenser fan motor is not operating, which of the following conditions could exist?
   a. the CFMT is open
   b. the ambient temperature is low
   c. all of the above
   d. none of the above

6. Which of the following components would open when the LLS is de-energized?
   a. HPS
   b. LPS
   c. CMT
   d. T

7. What is the purpose of the DT?
   a. to prevent the temperature of the freezer from rising to high on the defrost cycle
   b. to prevent the compressor from operating on the defrost cycle
   c. to operate the EFM on the defrost cycle
   d. to start the DTM

8. Which of the following components are connected in parallel?
   a. C and CFM
   b. CC and EFM
   c. LLS and DH
   d. all of the above

9. Write an operational sequence for both the defrost and freezing mode of the commercial freezer in Figure 6.9.

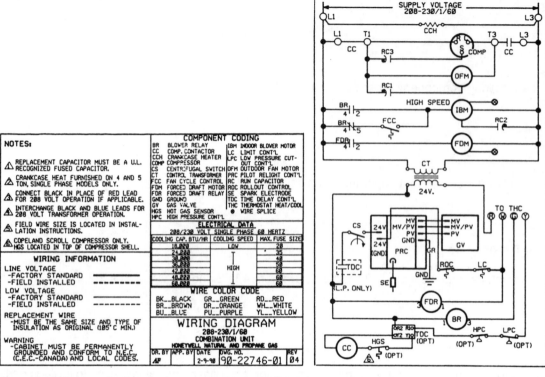

Figure 6.10 Schematic diagram for an electric air conditioning and gas heating unit to be used with questions in Part D. *(Courtesy of Rheem Air Conditioning Division, Fort Smith, AK)*

☐ D. Electric Air-Conditioning and Gas Heating Unit

Answer the following questions using Figure 6.10.

1. The centrifugal switch is located in the forced draft motor; what is its purpose in the control circuit?

2. The thermostat closes, making an electrical connection between R and W. What two safety controls could interrupt the low-voltage supply to the PRC?

3. The thermostat operates the IBM during the cooling cycle by making an electrical connection between R and G, but no provision is made for the thermostat to operate the IBM. How is the IBM energized during the heating cycle?

4. What controls the compressor contactor?

5. What is the supply voltage and control voltage of the unit?

6. Explain the operation of the crankcase heater.

7. True or False. The FDR and PRC are connected in series.

8. True or False. The HPC and LPC are connected in series.

9. True or False. The COMP and OFM are connected in parallel.

10. Write an operational sequence for the unit in Figure 6.10.

11. Have your instructor check your answers to questions 1–10.

☐ **E. Heat Pump**

Answer the following questions using Figures 6.11 and 6.12.

1. The transformer is located in _____.
   a. the outdoor unit   c. disconnect 1
   b. the indoor unit   d. none of the above

2. What does the first-stage cooling thermostat energize?
   a. MS   c. SC
   b. F   d. DFT

3. What does the second-stage cooling thermostat energize?
   a. MS   d. DFT
   b. F   e. a and b
   c. SC

4. What does first-stage heating thermostat energize?
   a. SC   c. F
   b. MS   d. b and c

5. What does the second-stage heating thermostat energize?
   a. SC   d. AH
   b. MS   e. c and d
   c. F

6. When the RHS-1 and RHS-2 switches are moved to the emergency heat position, which components are energized?
   a. AH   c. SC
   b. MS   d. compressor

7. CR-A is energized by the _____.
   a. MS   c. F
   b. SC   d. closing of disconnect 1

8. In the defrost cycle, the outdoor fan motor is _____.
   a. energized   b. de-energized

9. The defrost cycle is terminated by _____.
   a. DFT contacts (3 and 5)
   b. DT
   c. DFT contacts (3 and 5) or DT
   d. ODS

10. Write an operational sequence for the heat pump in Figure 6.11.

11. Have your instructor check your answers to questions 1–10.

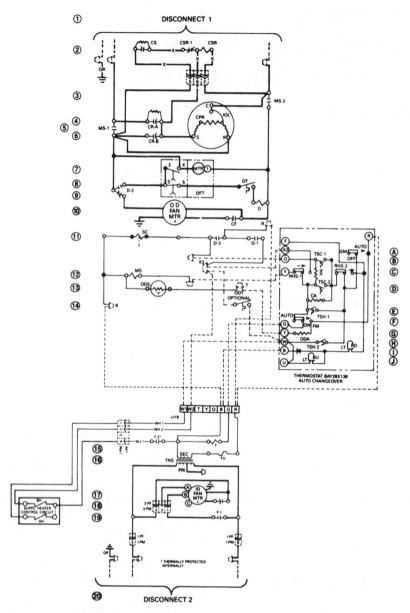

Figure 6.11 Complete schematic diagram of a heat pump to be used in questions in Part E. *(Courtesy of The Trane Company)*

Legend

AH:	Supplementary Heat Contactor	(19)		LVTB:	Low-voltage Terminal Board	
BH:	Supplementary Heat Contactor	(18)		MS:	Compressor Motor Contactor	(5, 3, & 12)
CA:	Cooling Anticipator			MTR:	Motor	
CR:	Run Capacitor	(4 & 6)		ODA:	Outdoor Temperature Anticipator	(G)
CPR:	Compressor			ODS:	Outdoor Temperature Sensor	(13)
D:	Defrost Relay	(9)		ODT:	Outdoor Thermostat	
DFT:	Defrost Timer	(7)		RHS:	Resistance Heat Switch	(C)
DT:	Defrost Termination Thermostat	(8)		SC:	Switchover Valve Solenoid	(11)
F:	Indoor Fan Relay	(15)		SM:	System Switch	(A)
FM:	Manual Fan Switch	(F)		TNS:	Transformer	(16)
HA:	Heating Anticipator			TSC:	Cooling Thermostat	(B & D)
HTR:	Heater			TSH:	Heating Thermostat	(E & H)
IOL:	Internal Overload Protection					
LT:	Light					

Figure 6.12 Legend for Figure 6.11.

SUMMARY STATEMENT: Describe the layout of a schematic diagram. Explain how a service technician would use the schematic on a service call.

Questions

1. A schematic diagram shows the components in _____.
   a. an energized position   b. a de-energized position

2. A schematic diagram is _____.
   a. an exact picture of a control panel with connecting wire
   b. an installation diagram
   c. a circuit-by-circuit layout of the control panel
   d. none of the above

3. A schematic diagram is most important for _____.
   a. servicing air-conditioning control systems
   b. locating components in the control panel
   c. installing air-conditioning equipment
   d. the homeowner

4. Most loads shown in schematic diagrams are connected in _____.
   a. series   b. parallel

5. Most switches shown in schematic diagrams are connected in _____.
   a. series   b. parallel

6. The thermostat used on the heat pump shown in Figure 6.11 is a _____ thermostat.
   a. single-stage heating   c. two-stage heating, single-stage cooling
   b. single-stage cooling   d. two-stage heating, two-stage cooling

7. If 24 volts are delivered to the MS coil in Figure 6.11, what action can be expected?
   a. MS contacts close, starting the compressor
   b. MS contacts close, starting the outdoor fan motor
   c. MS contacts close, starting the compressor and outdoor fan motor
   d. F-1 contacts close, starting the indoor fan motor

8. The defrost thermostat _____.
   a. opens on a temperature rise
   b. closes on a temperature rise

9. What controls the IF in Figure 6.5?
   a. M   b. IFR

10. The defrost timer motor runs when _____.
    a. DFT contacts 3 and 5 are closed
    b. D coil is energized
    c. TS-1 is closed
    d. MS is energized

Name: _____   Date: _____   Grade: _____

Comments:

Objectives: Upon completion of this lab, you should be able to draw basic schematic wiring diagrams following specifications and complete the electrical connections for the wiring diagrams.

Introduction: There will be times when you will have to draw basic schematic diagrams for control systems or circuits that you will be called on to install. You must keep in mind that a schematic diagram is laid out in a circuit-by-circuit arrangement; if this principle is followed, drawing of basic schematics can be easy.

Text References: Paragraphs 4.1 through 4.9

Tools and Materials: The following materials and equipment will be needed to complete this lab exercise.

Small screwdriver	Thermostat
Wire cutting pliers	Cleat receptacles
Wire strippers	Light bulbs to simulate loads
Terminal crimpers	Wire to make electrical connections
Supply power cords	Wire terminals (push-on and spade)
Transformer	12" × 18" piece of ½" plywood
Relays and contactors	Miscellaneous screws to attach components to plywood

Safety Precautions: Make certain all electrical components are securely attached to the plywood to prevent movement and possible electrical shock or short circuits. Allow your instructor to check your work before energizing a circuit or circuits. When a circuit or circuits are energized, make certain that electrical wiring connections are not touched; this could result in personnel injury or damage to equipment.

LABORATORY SEQUENCE (mark each box upon completion of task)

Draw the following diagrams and install components on a 12" × 18" piece of ½" plywood.

A. Diagram 1

☐ 1. Draw a wiring diagram with a single-pole switch controlling a 24-volt relay. The single-pole–single-throw contacts of the relay should operate a 115-volt light bulb. A transformer is required.

☐   2. Have your instructor check your wiring diagram.

☐   3. Make a list of the materials needed to complete your electrical circuit.

☐   4. Obtain the necessary electrical components from the supply room.

☐   5. Attach components securely to the plywood using screws.

☐   6. Make the necessary electrical connections to the components in order for the light bulb to be turned on when the switch controlling the relay is closed.

☐   7. Have your instructor check your circuit.

☐   8. Operate your electrical circuit by closing the switch.

☐   9. Observe the operation of the circuit. The light should burn.

☐   10. Disassembly of the board is not necessary at this time because many of the same components will be used in the next circuit.

B. Diagram 2

☐   1. Draw a wiring diagram with a single-pole switch controlling a 24-volt relay. The normally open set of contacts of the single-pole–double-throw relay should control one light bulb. The normally closed set of contacts of the single-pole–double-throw relay contacts should control the second light bulb. A transformer is required in this circuit.

☐   2. Have your instructor check your wiring diagram.

☐   3. Make a list of the materials needed to complete your electrical circuit.

☐   4. Obtain the necessary electrical components from the supply room.

☐   5. Attach components securely to the plywood using screws.

☐   6. Make the necessary electrical connections to the components in order for one light bulb to be on when the relay is de-energized and off when the relay is energized. The other light bulb will be off when the relay is de-energized and on when the relay is energized.

☐   7. Have your instructor check your circuit.

☐   8. Operate your electrical circuit by closing the switch.

☐   9. Observe the operation of the circuit.

☐ 10. Remove all components from the board and return them to the supply room.

C. Diagram 3

☐   1. Draw a wiring diagram using a thermostat to control four light bulbs representing the various functions of an air-conditioning system. The thermostat will be low voltage with the following letter designations: R representing the power source from the transformer, G representing the fan function, Y representing the cooling function, and W representing the heating function. Each system function will require a 24-volt relay to energize the light bulb representing that function. A 115-volt input to the 24-volt output transformer will be required for this control circuit.

☐   2. Have your instructor check your wiring diagram.

☐   3. Make a list of materials needed to complete your electrical circuit.

☐   4. Obtain the necessary components from the supply room.

☐   5. Attach components securely to the plywood using screws.

☐   6. Make the necessary connections to the electrical components in order for each light bulb to be turned on when that system function is called for by the thermostat.

☐   7. Have your instructor check your wiring.

☐   8. Operate your control system by selecting all modes of operation on the thermostat.

☐   9. Observe the operation of your control system.

☐ 10. Remove all components from the board and return them to the supply room.

D. Diagram 4

☐   1. Draw a schematic wiring diagram to meet the following specifications.
   a. Small packaged air conditioner
   b. Compressor with internal overload
   c. Two-speed evaporator fan motor
   d. Single-speed condenser fan motor
   e. 24-volt control system
   f. Low-voltage cooling thermostat
   g. Safety controls
      1) Low-pressure switch
      2) High-pressure switch
      3) Discharge line thermostat

☐   2. Have your instructor check your wiring diagram.

E. Diagram 5

☐   1. Draw a wiring diagram to meet the following specifications.
   a. Gas furnace with electric air conditioning
   b. 24-volt control system
   c. 24-volt heating/cooling thermostat
   d. Furnace will be equipped with a 24-volt ignition module controlling the gas valve supplying gas to the furnace. The furnace should be equipped with a flame roll-out thermostat and a temperature limit switch that opens, interrupting power to the ignition module and cutting off the gas supply to the furnace.
   e. The blower motor in the furnace will operate on high speed in cooling and low speed in heating. The blower will be controlled by the normally open contacts of the indoor fan relay in cooling and by the normally closed contacts of the indoor fan relay and a temperature fan switch that closes on a rise in temperature.
   f. The condensing unit will house a fan, compressor, condenser, and controls for automatic operation. The compressor and condenser fan motor will be connected in series and controlled by the contactor. The outdoor unit will be equipped with a high-pressure switch (opens on a rise in pressure) and a low-pressure switch (opens on a decrease in pressure).

☐   2. Have your instructor check your wiring diagram.

SUMMARY STATEMENT: Why is it important for technicians to understand how to draw a wiring diagram?

Questions

1. What is the purpose of the transformer?

2. Is the transformer wired like a load or a switch? Explain your answer.

3. What is the purpose of a contactor or relay?

4. What are the two main electrical parts of a relay or contactor?

5. Is a contactor coil wired like a load or a switch? Explain your answer.

6. Are contactor contacts wired like a load or a switch? Explain your answer.

7. Why are safety controls wired in series?

8. Why are loads wired in parallel?

9. What is the basic difference between an operating control and a safety control?

10. Why is it important for a service technician to know the function of a pressure switch in the circuit?

CHAPTER 7 | Alternating Current, Power Distribution, and Voltage Systems

Chapter Overview

Alternating current (AC) and direct current (DC) are two types of current that are used in the heating, cooling, and refrigeration industry today. Direct current is electron flow in only one direction and is used in most electronic control systems. Alternating current reverses the flow of electrons at regular intervals. Alternating current is used in most heating, cooling, and refrigeration equipment because its properties give it greater flexibility than direct current. Alternating current is easier to transmit over a long distance and no expensive transmission equipment is required. It also has the capability of supplying almost any voltage that is needed by the customer.

Electron flow in an alternating current circuit changes direction twice per cycle. The number of cycles per second of alternating current is the frequency (usually 60 cycles per second).

Most electric utilities supply alternating current to their customers. Alternating current is produced by an alternator. When direct current is needed, it can easily be produced by a direct current generator or a rectifier. When needed, direct current is usually produced by the customer. Automobile electrical systems are direct current systems that use a battery or vehicle to produce the power supply.

Direct current has a limited use in the industry; it is used mostly for refrigeration transport equipment, electronic air cleaners, electronic control components, and electronic control systems. Alternating current is used as the basic source to supply power to HVAC equipment and most electric control systems. Because of the popularity of alternating current in the HVAC industry, it is important for industry personnel to be familiar with AC theory, power distribution, and common voltage systems.

In an AC circuit, the voltage leads the current due to inductive reactance or the current leads the voltage due to capacitive reactance. The reactance is the resistance an AC circuit encounters when the flow of electrons is reversed. The total resistance of an AC circuit is called the *impedance* and is the sum of the reactance and resistance of the circuit. Voltage and current are out of phase in an AC circuit. Consequently, the power of the circuit must be calculated by using the effective wattage. The ratio between the true power and the effective power in an AC circuit is called the *power factor*.

Electric utilities can supply single-phase or three-phase power to structures at voltages from 208 volts to 460 volts. The generating plant supplies a high voltage that is stepped down, making it usable to the customer. Common voltage characteristics are 230-V–1-phase–60-Hz, 208-V–3-phase–60-Hz, 230-V–3-phase–60-Hz, and 460-V–3-phase–60-Hz. Most residences are supplied with single-phase current while most commercial and industrial applications are supplied with three-phase current. Single-phase current supplies two hot legs of power while three-phase current is supplied with three hot legs. Many structures are being supplied with high-voltage systems, which are becoming increasingly popular. Technicians must be familiar with the many different voltage systems that they will encounter in the industry.

Key Terms

Alternating current	Frequency	Reactance
Alternator	Impedance	Sine wave
Capacitive reactance	Inductance	Single phase
Delta system	Inductive reactance	Three phase
Direct current	Peak voltage	Wye system
Effective voltage	Power factor	

REVIEW TEST

Name: _____   Date: _____   Grade: _____

Match the following terms and definitions.

_____ 1. Direct current

a. 60 Hertz

_____ 2. Alternating current

b. Sum of resistance and reactance in an AC circuit

_____ 3. Sine wave

c. Reached at a peak of 90 electrical degrees

_____ 4. Frequency

d. 230-V–3ø–60-Hz

_____ 5. Peak voltage

e. Electron flow in one direction

_____ 6. Effective voltage

f. Resistance to flow reversals in alternating current circuit

_____ 7. Phase

g. Number of currents alternating at different time intervals

_____ 8. Alternator

h. Producer of alternating current

_____ 9. Inductance

i. A graphical representation of alternating current

_____ 10. Reactance

j. 208-V–3ø–60-Hz

_____ 11. Inductive reactance

k. Electrons reversing their flow of direction at regular intervals

_____ 12. Capacitive reactance

l. Frequently used electrical service for residences

_____ 13. Impedance

m. 0.707 times peak voltage

_____ 14. Power factor

n. Number of complete cycles that occur in a second

_____ 15. Single phase

o. Induced voltage that counteracts initial voltage

_____ 16. Three phase

p. Caused in an AC circuit by using capacitors

_____ 17. Delta system

q. Produces an out-of-phase condition between the voltage and amperage in a circuit

_____ 18. Wye system

r. True power divided by apparent power

_____ 19. 60 Hz

s. Power supply of three hot legs of power

_____ 20. 230-V–1ø–60-Hz

t. Power supply of two hot legs of power

CHAPTER 8 Installation of Heating, Cooling, and Refrigeration Systems

Chapter Overview

The proper installation of heating, cooling, and refrigeration equipment is as important as any other phase of the industry. Installation covers a broad range of subjects, but one of the most important is the electric circuit servicing the equipment. It is the responsibility of the installation technician to insure that the correct size conductor is supplying power to the unit. On many occasions, the service technician is required to check the electrical connections to the unit to make certain that troubles have not occurred in that area. One of the most important parts of an equipment installation are the electrical connections made to the unit, both power and control. This chapter is concerned with the installation of the power supply to the equipment. Equipment manufacturers supply a wire sizing chart to determine the correct conductor that should be used to supply electrical power to the equipment. The correct wire size can be easily obtained from the table of recommended wire sizes from the installation instructions. The figures that are used in this table will correspond to the *National Electrical Code®*. The *National Electrical Code®* governs the type and sizes of wire that can be used for a particular application and a certain amperage. The wire sizing chart from the *National Electrical Code®* should be used if there is any doubt about the conductor supplying a specific load.

Copper is the most popular conductor used in the industry to supply electrical power to HVAC equipment. In some rare cases, aluminum is used as a conductor. Copper wire is an excellent conductor of electricity and has many other characteristics that make it the conductor of choice for manufacturers and installers in the industry. Factors to be considered when sizing circuit conductors are voltage drop, insulation type, enclosure used, and safety. Wire sizing charts are available in the *National Electrical Code®* and should be used when questions arise. The technician should make certain that the voltage drop in any circuit will maintain minimum standards.

All heating, cooling, and refrigeration equipment should have some means of disconnecting the power supply at the equipment. Some equipment has a built-in means for disconnecting the power, such as circuit breakers or fuse blocks. In most cases, a means of disconnecting power would have to be supplied and installed by the technician. If any doubt is raised about the installation of the disconnect, refer to state and local codes. The proper type of disconnect should be used to suit the application.

In most cases, power will be supplied from a fusible load center, breaker panel, or distribution center to the disconnect means supplying the equipment. The fusible load center is an electrical panel that was used in the past, but many still exist in the field and in many cases technicians will be required to obtain electrical energy for a piece of equipment from this type of panel. The breaker panel is the most popular panel used in residences and larger applications. However, many technicians will be required to use the larger distribution panels that are popular in commercial and industrial applications.

In many cases, the installation technician will be required to make electrical connections from the source to the equipment. The technician should make certain to follow the *National Electrical Code®*.

Key Terms

American Wire Gauge	**Disconnect switches**	**Insulators**
Breaker panel	**Distribution center**	***National Electrical Code®***
Breakers	**Fusible load center**	**Nonfusible switches**
Conductors	**Fusible switches**	

REVIEW TEST

Answer the following questions.

1. What organization governs the types and sizes of wire that can be used for a particular application?

2. How are conductors larger than 4/0 sized?

3. What would be the best conductor to use when placed in dry and wet locations with a temperature of less than 140°F?

4. What would be the voltage drop of a 750′ electrical circuit supplied with 10 TW wire and a load of 42 amps?

5. What factors should be considered when choosing a conductor for an electrical circuit?

6. What is the purpose of a disconnect switch?

7. What is the difference between a light-duty disconnect switch and a heavy-duty disconnect switch?

8. Where there is a danger of explosion, what type of disconnect would have to be used?

9. What size fuses could be used in a 30A, 60A, 100A, and 200A disconnect?

10. What is a fusible load center?

11. What does the term "pole" mean in reference to a circuit breaker?

12. If a 115-volt circuit is needed and there is no space in a breaker panel, how could the technician remedy the situation?

13. Where are distribution centers used?

14. What size conductor would be required for a packaged unit with a compressor rated at 35 FLA, a condenser fan motor rated at 06 FLA, and an evaporator fan motor rated at 10 FLA?

15. What is the purpose of a main breaker?

16. Why would a technician use a nonfusible disconnect instead of a fusible disconnect?

17. What are the responsibilities of an installation technician as far as the power supply to the equipment is concerned?

18. What type of circuit breaker would be used to supply power to a single-phase condensing unit?

19. Why is it important for a service technician to know how to size wire for a condensing unit?

20. How can a service technician determine the condition of a circuit breaker?

LAB 8–1 Sizing Wire for an Air-Conditioning Condensing Unit

Name: _____	Date: _____	Grade: _____

Comments:

Objectives: Upon completion of this lab, you should be able to correctly size, install, and make the proper electrical connections of the conductors to an air-cooled condensing unit in the lab.

Introduction: One of the most important tasks that technicians perform in the field is installing HVAC equipment. Proper installation is essential for correct and efficient operation of new equipment. The technician must be able to select the correct size conductor for a particular application and install the conductor in order to supply electrical energy to the equipment.

Text Reference: Chapter 8

Tools and Materials: The following materials and equipment will be needed to complete this lab exercise.

National Electrical Code®
Installation instructions of unit assigned by instructor
Air-conditioning condensing unit
Wire and supplies to install power wiring to unit
Miscellaneous electrical supplies
Electrical meters
Wire cutting pliers
Wire strippers
Screwdrivers
Hacksaw

Safety Precautions: Make certain that the electrical source is disconnected when making electrical connections. In addition,

- Make sure all connections are tight.
- Make sure no current-carrying conductors are touching metal surfaces except the grounding conductor.
- Make sure the correct voltage is being supplied to the unit.
- Make sure all covers on the equipment are replaced.
- Make sure body parts do not come in contact with live electrical conductors.
- Keep hands and materials away from moving parts.

LABORATORY SEQUENCE (mark each box upon completion of task)

A. Sizing Wire for a Condensing Unit Using the Manufacturer's Installation Instructions

☐ 1. See instructor for assignment of air-conditioning condensing unit. The control wiring for the condensing unit should be installed.

☐ 2. Read the section in the installation instructions on wiring the unit.

☐   3. Follow the procedures described in the installation instructions to determine the correct wire size for the assigned unit. Take note of the distance between the condensing unit and the electrical supply panel. (NOTE: The type of insulation on the conductor plays an important part in its current-carrying ability.)

☐   4. Complete Data Sheet 8A.

DATA SHEET 8A

Condensing unit model number _____

Condensing unit serial number _____

Distance from power supply to unit _____

Size of wire to be installed _____

List of materials required to complete the installation:

☐   5. Have your instructor check your wire sizing and materials list for the assigned condensing unit.

☐   6. Make the necessary electrical connections to the condensing unit using the correct size conductor assigned by the instructor.

☐   7. Operate the unit and check the amperage of the following components.

Condensing unit _____

Compressor _____

Condenser fan motor _____

☐   8. Clean up the work area and return all tools and supplies to their correct location(s).

B. Sizing Wire for a Condensing Unit Using the *National Electrical Code*®

☐   1. See your instructor for assignment of air-conditioning condensing unit.

☐   2. Using the *National Electrical Code*®, size the power wiring conductors supplying power to the condensing unit. (NOTE: Refer to paragraph 8.1 if you need help.)

☐   3. Calculate the voltage drop in the conductor from the unit to the panel.

Voltage drop _____

☐   4. Complete Data Sheet 8B.

Condensing unit model number_____

Condensing unit serial number _____

Distance from power supply to unit _____

Size of wire to be installed_____

Voltage drop of circuit _____

List of materials required to complete the installation:

☐   5. Have your instructor check your wire sizing and materials list for the assigned unit.

MAINTENANCE OF WORK STATION AND TOOLS: Clean and return all tools to their proper location(s). Replace all equipment covers. Clean up the work area.

SUMMARY STATEMENT: Why is it important to properly size the conductor supplying electrical power to air-conditioning equipment? What may happen if the conductor is too small?

Questions

1. According to the *National Electrical Code*®, what size wire has an ampacity of 65 amps with THWN insulation?

2. What is the size and insulation type of the conductor needed to supply a condensing unit whose compressor draws 36 amps and whose condenser fan motor draws 5 amps?

3. When no state or local codes are in force, what regulations should be followed when making electrical connections?

4. Why is copper the most popular conductor used in the HVAC industry?

5. Why is it important to check the voltage drop on extremely long electrical circuits?

6. What type of circuit breaker would be used on the following applications?

    Three-phase motor:

    Single-phase condensing unit:

    120-volt fan motor:

7. How are breakers installed in a breaker panel?

8. What factors must be considered when sizing conductors for a condensing unit?

9. When installing a conductor for a condensing unit, what electrical checks should be made upon completion of the installation?

10. What type of disconnect switch would be used for a condensing unit when mounted outdoors?

CHAPTER 9 | Basic Electric Motors

Chapter Overview

The electric motor is the most important load in a refrigeration system. The electric motor changes electrical energy into mechanical energy. Regardless of the application or the type of equipment, electric motors play an important part in the industry. Motors are used to drive compressors, fans, pumps, dampers, and any other device that needs energy to power its movement. There are many different types of electric motors, each with different running and starting characteristics. Most single-phase motors are designed and used according to their running and starting torque. In an electric motor, electrical energy is changed to mechanical energy by magnetism, which causes the motor to rotate. A rotating magnetic field is responsible for the continuous rotation of an electric motor.

The HVAC industry uses all kinds of AC motors to rotate the many different devices that require rotation in a complete system. There are many kinds of AC motors because of the many different tasks that are required of motors in the industry. Different applications require different starting and running characteristics. For example, some compressors require a motor with high starting torque and good running efficiency, while small propeller fans use a motor with low starting torque and average running efficiency. Electric motors are selected primarily for their starting torque. The five basic types of AC motors used in the industry (listed with the lowest starting torque first) are shaded pole, split phase, permanent split capacitor, capacitor-start–capacitor-run, and three phase. The shaded-pole motor has low starting torque and is used on devices that start with very little resistance, such as propeller fans, small furnace fan motors, and small condenser fan motors. These motors can be easily stalled but because of the small locked rotor amp draw can stop and still not burn out the motor winding. Some shaded-pole motors can be mechanically reversed.

Capacitors are devices that are used with electric motors to increase their starting torque by producing a second phase of electrical current. A capacitor stores electrons momentarily. The two types of capacitors used in the industry are the starting and running capacitors. The starting capacitor is placed in a motor circuit for only a short period of time to increase the starting torque of the motor. Once the motor has reach approximately 75% of full speed, the capacitor is dropped out of the circuit. A running capacitor is designed to remain in the electrical circuit while the motor is running. The running capacitor increases the motor's starting torque and running efficiency. The running capacitor is oil filled to prevent overheating. Capacitors are rated in microfarads. Refer to the text for capacitor rules that can be used as a guide for the replacement of capacitors. Capacitors can be checked with an ohmmeter or commercial capacitor tester.

Two general classifications of split-phase motors are used in the industry. The resistance-start–induction-run

motor and the capacitor-start–induction-run motor are the types of split-phase motors that are in common use today. The split-phase motor has a start and run winding and drops the starting winding out of the electrical circuit by a centrifugal switch once the motor reaches 75% of its rated speed. The capacitor-start motor uses a capacitor to increase the starting torque of the motor and, like the split-phase motor, the capacitor and starting winding must be removed from the circuit after the motor has attained 75% of its operating speed. In a capacitor-start motor, the starting winding and starting capacitor are dropped from the circuit by a centrifugal switch on an open-type motor and some type of starting relay on an enclosed motor. Split-phase and capacitor-start motors are used on belt-drive devices such as fans and air compressors. These motors can usually be reversed electrically.

Permanent split-capacitor (PSC) motors are simple in design with a moderate starting torque and good running efficiency, which makes them very popular. The PSC motor has both starting and running windings. PSC motors are used on compressors, where refrigerant equalizes on the off cycle, on direct-drive fan motors, and in many other applications. The PSC motor is not reversible unless it has been designed with this feature. Many of these motors that are easily reversed electrically are being manufactured.

The capacitor-start–capacitor-run motor produces a high starting torque and good running efficiency. The starting winding is energized at all times with the starting capacitor being removed from the circuit. This type of motor is used in compressors and other applications that require a high starting torque and good running efficiency. A starting relay is used to drop the capacitor out of the circuit.

The three-phase motor is supplied by three phases of electrical power. These motors have a very high starting torque without any additional components. Three-phase motors are used in most commercial applications where the structure's electrical service is three phase. A three-phase motor is easily reversible and (if so equipped) can be operated on either high or low voltage.

The electronically commutated motor, ECM, is a high-efficiency, variable-speed motor that is interfaced with system controls. The ECM can be programmed to maintain constant air flow over a wide range of external static pressures. The ECM is constructed as a three-phase DC motor. The motor is made in two parts: One part is the motor and the other section is the controller. The motor can be disconnected and checked just like a three-phase motor. The controller/module can easily be checked with a motor analyzer purchased from the manufacturer.

The most common type of enclosed motor is the hermetic compressor motor. Hermetic compressor motors are completely sealed in a hermetic casing with the compressor. Enclosed motor connections are made from the motor to the outside of the enclosure by terminals that extend outside of the casing. All starting components are connected outside of the hermetic compressor. All single-phase hermetic compressor motors have a start and run winding; service technicians must be able to determine the common, start, and run terminals of a single-phase hermetic compressor. Three-phase hermetic compressors have three electrical windings and require no starting components. Three-phase hermetic compressors that are dual-voltage models will have more than three terminals on the compressor.

Troubleshooting electric motors can generally be divided into two categories, electrical problems and mechanical problems. Electrical problems are those with the electrical continuity of the motor windings or their starting components. Mechanical problems are those that are related to the mechanical parts of the motor such as bad bearings. The technician must be able to correctly diagnose problems that exist with the motor or motors in an HVAC system.

Key Terms

Capacitor
Capacitor-start motor
Capacitor-start–capacitor-run motor
Electronically commutated motor (ECM)
Electromagnet
Hermetic compressor

Induced magnetism
Magnetic field
Magnetism
Microfarad
Permanent split-capacitor motor
Rotor
Running capacitor

Shaded-pole motor
Split-phase motor
Starting capacitor
Stator
Three-phase motor
Torque

REVIEW TEST

Name: _____ Date: _____ Grade: _____

Complete the following multiple choice questions by selecting the correct answer.

1. Torque is the _____.
   a. starting power of an electric motor
   b. thrust of an electric motor
   c. rotation of an electric motor
   d. none of the above

2. A two-pole, 60-Hz motor rotates at a speed of _____.
   a. 1100 rpm
   b. 1750 rpm
   c. 3500 rpm
   d. 7200 rpm

3. Which of the following is not a common electric motor used in the HVAC industry?
   a. shaded pole
   b. PSC
   c. three phase
   d. torque master

4. The rotation of the shaded-pole motor is _____.
   a. away from the shaded pole
   b. toward the shaded pole
   c. determined by the electrical connection
   d. always the same

5. How can a shaded-pole motor be reversed?
   a. reverse the electrical connections
   b. mechanically by reversing the rotor
   c. mechanically by reversing the stator
   d. reverse the power supply

6. The unit of measurement for the strength of a capacitor is the _____.
   a. volt
   b. farad
   c. microfarad
   d. amp

7. A capacitor _____.
   a. stores electrons momentarily
   b. produces electrons
   c. stores electrons permanently
   d. changes the polarity of electrons

8. A running capacitor _____.
   a. increases starting torque
   b. increases running efficiency
   c. is oil filled to dissipate heat
   d. all of the above

9. The voltage of a replacement capacitor must be _____.
   a. equal to the voltage of the original capacitor
   b. greater than the voltage of the original capacitor
   c. equal to or greater than the original capacitor
   d. exactly the same as the original capacitor

10. What would be the strength of two 15-microfarad running capacitors connected in parallel?
   a. 7.5 microfarads
   b. 15 microfarads
   c. 30 microfarads
   d. 225 microfarads

11. Which of the following capacitors could be used to replace a 145-microfarad, 250-volt start capacitor?
   a. 125 microfarad, 250 volts
   b. 145 microfarad, 115 volts
   c. 165 microfarad, 330 volts
   d. 245 microfarad, 250 volts

12. What usually drops the starting winding out of the circuit in a split-phase, open-type motor?
   a. centrifugal switch
   b. manual switch
   c. current-type relay
   d. potential relay

13. What usually drops the starting winding and capacitor out of the circuit in a hermetic capacitor-start motor?
   a. centrifugal switch
   b. manual switch
   c. current-type relay
   d. none of the above

14. What type of motor is commonly used on compressors where the refrigerant equalizes during the off cycle and in direct-drive fan motor applications?
   a. capacitor start
   b. permanent split capacitor
   c. shaded pole
   d. three phase

15. The ECM is a
   a. three-phase motor
   b. single-phase motor
   c. three-phase DC motor
   d. shaded-pole motor

16. On a three-phase electric motor, each winding is approximately _____ electrical degrees out of phase with the others.
   a. 90
   b. 120
   c. 180
   d. 360

17. If a service technician measured 50 ohms between the compressor start winding and the compressor housing, what would be the condition of the compressor?
   a. shorted winding
   b. open winding
   c. grounded winding
   d. good winding

18. What electrical meter would a service technician use to check the electrical condition of an electric motor winding?
   a. voltmeter
   b. ammeter
   c. ohmmeter
   d. wattmeter

19. What are the possible problem areas in an open capacitor-start motor?
   a. windings, bearings, centrifugal switch, and capacitor
   b. windings and bearings
   c. windings, bearings, and capacitor
   d. windings, centrifugal switch, and capacitor

20. A service technician checks a hermetic compressor. The compressor is hot to the touch and measures infinite resistance between the common to start and run windings. What would be the possible cause of the problem?
   a. bad start winding
   b. bad run winding
   c. open internal overload
   d. none of the above

LAB 9–1 Identifying Types of Electric Motors

Name: _____	Date: _____	Grade: _____

Comments:

Objectives: Upon completion of this lab, you should be able to identify the most common types of motors used in the HVAC industry.

Introduction: The technician will often have to replace electric motors when repairing HVAC equipment. All electric motors are not the same but are designed to meet specific applications. It is important for the technician to know what type of electric motor is being replaced so that the correct replacement can be installed. There are six general types of electric motors commonly used in the industry:

1. Shaded pole
2. Split phase
3. Permanent split capacitor
4. Capacitor-start–capacitor-run
5. Three phase
6. Electronically commutated motor (ECM)

Often, the best source of information about a motor is the nameplate if it is still readable or the unit wiring diagram; on other occasions, the technician will be required to determine the type of motor being replaced just by examination of the faulty motor.

Text Reference: Chapter 9

Tools and Materials: The following materials and equipment will be needed to complete this lab exercise.
Selection of electric motors installed in equipment and removed from equipment
Motor manufacturers' catalogs
Wire cutting pliers
Nut drivers
Screwdrivers

Safety Precautions: When examining motors installed in equipment, make certain that the electrical power source is disconnected.

LABORATORY SEQUENCE (mark each box upon completion of task)

A. Determining the Types of Motors Used for Specific Applications

Several things can help you identify the types of motors:

- Examine the motor and nameplate.
- Shaded-pole motors have a shading coil in the stator that can easily be identified.
- Permanent split-capacitor motors can be identified because their windings look like a maze of wires.
- Split-phase and capacitor-start motors can have an open design that is mounted on a frame, or they can be sealed hermetic compressor motors. If the motor is a capacitor-start motor, a capacitor will usually be mounted on top of the motor. The hermetic compressor motor is almost impossible to identify without the manufacturer's information. Capacitor-start–capacitor-run motors are mainly used in hermetic compressors. An open-type capacitor-start–capacitor-run motor will have two capacitors (one run and one start) mounted on the motor.

☐ 1. See your instructor for assignment of five electric motors in equipment.

2. Examine the five electric motors.

3. Examine the wiring diagram(s) and any other available information about the equipment and the motors.

4. Record each type of motor in Data Sheet 9A.

5. Have your instructor check Data Sheet 9A.

DATA SHEET 9A

Motor	Type of Motor
#1	_____
#2	_____
#3	_____
#4	_____
#5	_____

B. Determining the Types of Five Electric Motors

1. See your instructor for assignment of five electric motors in the lab.

2. Examine the five electric motors assigned by the instructor.
   NOTES: • Utilize nameplate if available and readable.
   • Look for motor items such as: capacitors, switches, frame, shaded poles, etc.
   • Refer to any motor manufacturers' catalogs for information.

3. Make the best possible judgment about the assigned motors and complete Data Sheet 9B.

4. Have your instructor check Data Sheet 9B.

DATA SHEET 9B

Motor	Type of Motor
#1	_____
#2	_____
#3	_____
#4	_____
#5	_____

MAINTENANCE OF WORK STATION AND TOOLS: Clean and return all tools to their proper location(s). Replace all equipment covers. Clean up the work area.

SUMMARY STATEMENT: Why is it important for the HVAC technician to know the type of motor that is being replaced?

Questions

1. What type of motor would be used for a hermetic compressor with the mechanical refrigeration system equalizing on the off cycle?

2. Why is it difficult to determine the type of motor if no information is available except examination?

3. How would you identify an open-type capacitor-start motor?

4. Why are three-phase motors easier to identify than single-phase motors?

5. What is the advantage when the motor being replaced has a nameplate?

6. What type of motor would be used for a small propeller fan?

7. How would the technician determine if a hermetic compressor motor was a PSC or CSR motor?

8. What are some key facts that a technician could use in identifying PSC motors?

9. How could a visual inspection differentiate between an open split-phase motor and an open capacitor-start motor?

10. What would be your process for determining the type of a specific motor if only a visual inspection can be made?

LAB 9–2 Capacitors

Name: _____   Date: _____   Grade: _____

Comments:

Objectives: Upon completion of this lab, you should be able to determine the condition of starting and running capacitors. You will be able to use the capacitor rules to select replacement capacitors.

Introduction: Many electric motors used in the HVAC industry require a capacitor or capacitors for adequate starting torque and to increase running efficiency. The technician must be able to determine the condition of capacitors and, utilizing a set of capacitor rules, size and replace a capacitor with capacitors that are available from the service truck; often, this will allow the technician to make the necessary repairs without a trip to the local wholesaler.

Text Reference: Paragraph 9.5

Tools and Materials: The following materials and equipment will be needed to complete this lab exercise.
    10 numbered capacitors for identification (Capacitor Kit #1)
    10 numbered capacitors for checking (Capacitor Kit #2)
    10 numbered capacitors for using capacitor rules (Capacitor Kit #3)
    Analog volt-ohmmeter
    Capacitor tester
    2-watt, 15,000-ohm resistor
    Insulated needle nose pliers
    Screwdrivers

Safety Precautions: Make certain that all capacitors have been discharged (shorted) with the 2-watt 15,000-ohm resistor before touching the terminals to prevent electrical shock or possible damage to an electric meter. Capacitors often remain charged even while sitting on the shelf. Use insulated pliers.

LABORATORY SEQUENCE (mark each box upon completion of task)

A. Determining the Size and Type of Capacitor

☐   1. Obtain Capacitor Kit #1 from your instructor.

☐   2. Using a 2-watt, 15,000-ohm resistor, short between the terminals of the capacitors to discharge any capacitor that may be charged. (See Figure 9.1.)

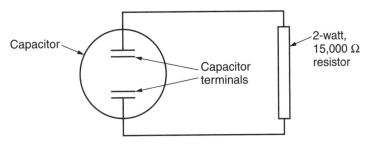

Figure 9.1 Discharging a capacitor with a resistor.

☐ 3. Record the types, voltages, and microfarad ratings of the capacitors in Capacitor Kit #1 on Data Sheet 9C.

DATA SHEET 9C

Capacitor #	Type	Voltage	Rating (µF)

☐ 4. Have your instructor check your data sheet.

☐ 5. Return Capacitor Kit #1 to its proper location.

B. Checking the Condition of Capacitors

☐ 1. Obtain Capacitor Kit #2 from your instructor.

☐ 2. Check the ten capacitors in Capacitor Kit #2 using the following procedure. Complete Data Sheet 9D as each capacitor is checked.

☐ 3. Using a 2-watt, 15,000-ohm resistor, short between the terminals of the capacitor to discharge any capacitor that may be charged.

☐ 4. Set the selector switch on the meter to the R × 100 scale.

☐ 5. Check each capacitor by touching one lead to each run capacitor terminal while watching the meter. The needle of the meter should rise and fall if the capacitor is good. The needle of the meter should fall back to infinity on the meter scale. If no needle movement takes place, reverse the meter leads. If the needle falls back only to a certain point, the capacitor has an internal short of the value shown on the meter. If you are checking a starting capacitor with a bleed resistor, the meter reading will fall back no further than the value of the bleed resistor. If using a digital meter, the meter will count up and back down. Record the condition of each capacitor on Data Sheet 9D.

DATA SHEET 9D

| Capacitor # | Type | CONDITION | |
		Voltage	Rating (μF)

☐ 6. Check the running capacitors for a grounded condition by setting the meter selector switch to the R × 1000 scale. Place one meter lead on the capacitor terminals and the other lead on the capacitor case. If any needle movement occurs, the capacitor is shorted to the case and is defective. This test is performed only on running capacitors because the casing of a start capacitor is plastic.

☐ 7. Check the ten capacitors with a capacitor tester following the capacitor tester instructions. If you have any questions, check with your instructor. Record the condition and rating (if capacitor is so equipped) of the ten capacitors in the appropriate section of Data Sheet 9D.

8. Have your instructor check your data sheet.

9. Return Capacitor Kit #2 to its proper location.

C. Replacing Capacitors Using Capacitor Rules

1. Obtain Capacitor Kit #3 from your instructor.

2. Data Sheet 9E has a list of six capacitors (three run and three start) that are to be replaced using the capacitors in Capacitor Kit #3. Your instructor will provide the capacitance of each of the six capacitors. Record which of the capacitors in Capacitor Kit #3 is being used to replace each capacitor on the list. Use the following capacitor rules to select the proper capacitor. Often, technicians are unable to obtain the exact replacement capacitor but have the opportunity to replace a capacitor with a capacitor or capacitors that are available on the service truck. This exercise could require that more than one capacitor be used.

DATA SHEET 9E

Capacitor #	Type	Replacement Capacitor(s)
#1	Run	_____
#2	Run	_____
#3	Run	_____
#4	Start	_____
#5	Start	_____
#6	Start	_____

☐ 3. Capacitor rules:
    a. The voltage of any capacitor used for replacement must be equal to or greater than the capacitor being replaced.
    b. The strength of the starting capacitor replacement must be at least equal to but no greater than 20% of the capacitor being replaced.
    c. The strength of the running capacitor replacement may be plus or minus 10% of the capacitor being replaced.
    d. If capacitors are installed in parallel, the sum of the capacitors is the total capacitance.
    e. The total capacitance of capacitors in series may be found using the following formula:

$$C_T = \frac{C_1 \times C_2}{C_1 + C_2}$$

☐ 4. Complete Data Sheet 9E by listing the capacitors from Capacitor Kit #3 that could replace the six capacitors listed on the data sheet. More than one capacitor could be required.

☐ 5. Have your instructor check your data sheet.

☐ 6. Return Capacitor Kit #3 to the proper location.

MAINTENANCE OF WORK STATION AND TOOLS: Clean and return all tools to their proper location(s). Clean up the work area.

SUMMARY STATEMENT: What is the purpose of capacitors in electric motors?

Questions

1. What are the major differences between a starting capacitor and a running capacitor?

2. Explain the construction of a starting capacitor.

3. Explain the construction of a running capacitor.

4. What is the advantage of using a bleed resistor on a starting capacitor?

5. What would be the danger of using a running capacitor that is grounded (shorted) to the case?

6. What is the purpose of a run capacitor in a motor circuit?

7. Why is it important to know and be able to use the capacitor rules?

8. How would a faulty run capacitor affect the operation of a condenser fan motor?

9. What would be the symptoms of an open start capacitor on a capacitor-start motor?

10. What is done to the voltage of capacitors when they are connected in series?

LAB 9–3 Single-Phase Electric Motors

Name: _____   Date: _____   Grade: _____

Comments:

Objectives: Upon completion of this lab, you should be able to correctly make the electrical connections to shaded-pole, split-phase, capacitor-start, and permanent split-capacitor motors, operate each motor, and complete the associated data sheet.

Introduction: The service technician will be required to install or replace all types of single-phase motors in HVAC equipment. The installation of single-phase motors will include the electrical connections, direction of rotation, and final check to determine if the motor is operating properly.

Text References: Paragraphs 9.4, 9.6, and 9.7

Tools and Materials: The following materials and equipment will be needed to complete this lab exercise.
- Split-phase motor for disassembly
- Shaded-pole motor
- Split-phase motor
- Capacitor-start motor
- Permanent split-capacitor motor
- Power supply to operate motors
- Volt-ohmmeter
- Clamp-on ammeter
- Miscellaneous electrical handtools

Safety Precautions: Make certain that the electrical source is disconnected when making electrical connections. In addition:
- Make sure all electrical connections are tight.
- Make sure no bare conductors are touching metal surfaces except the grounding conductor and the connection terminals.
- Make sure the correct voltage is being supplied to the motor.
- Keep hands and other materials away from the rotating shaft.
- Make sure body parts do not come in contact with live electrical circuits.

LABORATORY SEQUENCE (mark each box upon completion of task)

A. Disassembling a Split-Phase Electric Motor

☐ 1. Obtain a split-phase motor from your instructor for disassembly.

☐ 2. Mark the end bell location on the winding section of the motor to insure correct placement when reassembled.

☐ 3. Remove bolts connecting the end bells and the winding section of the motor.

☐ 4. Carefully disassemble the motor.

☐ 5. Examine the end bells, centrifugal switch, rotor, start and run windings, and bearings.

☐ 6. Reassemble the motor, making sure the marks on the stator and end bells match.

B. Shaded-Pole Electric Motors

☐ 1. Obtain a two-speed shaded-pole motor from your instructor.

☐ 2. Check the resistance of the windings of the motor with a volt-ohmmeter, record the resistance readings, and draw a schematic diagram of the motor.

   Resistance readings _____

☐ 3. Make the proper electrical connections to the motor.

☐ 4. Have your instructor check your wiring and wiring diagram.

☐ 5. Operate the motor on both speeds and record its current draw.

   Current draw (Speed 1) _____

   Current draw (Speed 2) _____

☐ 6. Remove the electrical wiring from the motor and return the motor to the proper location.

☐ 7. Locate a shaded-pole motor on a unit in the lab and record its location.

   _____

C. Permanent Split-Capacitor Motors

☐ 1. Obtain an open two-speed PSC motor from your instructor.

☐ 2. Check the resistance of the windings of the motor with a volt-ohmmeter, record the resistance readings, and draw a schematic diagram of the motor.

   Resistance readings _____

☐ 3. Select and obtain the proper capacitor.

☐ 4. Make the proper electrical connections to the motor.

☐ 5. Have your instructor check your wiring and wiring diagram.

☐ 6. Operate the motor on both speeds and record its current draw.

Current draw (Speed 1) _____

Current draw (Speed 2) _____

☐ 7. Remove the electrical wiring from the motor and return the motor to the proper location.

☐ 8. Locate a PSC motor on a unit in the lab and record its location.

_____

D. Split-Phase and Capacitor-Start Motors

☐ 1. Obtain an open split-phase or capacitor-start motor from your instructor.

☐ 2. Check the resistance of the windings of the motor with a volt-ohmmeter, record the resistance readings, and draw a schematic diagram.

Resistance readings _____

Schematic of split-phase motor

☐ 3. Make the proper electrical connections.

☐ 4. Have your instructor check your wiring and wiring diagram.

☐ 5. Operate the motor and record its current draw.

Current draw _____

☐ 6. Reverse the motor.

☐ 7. Remove the electrical wiring from the motor and return the motor to the proper location.

☐ 8. Locate a split-phase or capacitor-start motor on a unit in the lab and record its location.

_____

MAINTENANCE OF WORK STATION AND TOOLS: Clean and return all tools to their proper location(s). Replace all equipment covers. Clean up the work area.

SUMMARY STATEMENT: Why is it important to use the correct type of single-phase motor for the application in which it is being used?

Questions

1. What would be the results of using a shaded-pole motor for a high starting torque application?

2. How can a shaded-pole motor be reversed? Can all shaded-pole motors be reversed?

3. The individual windings of a dual voltage (115/230) split-phase motor are designed for what voltage?

4. What is the difference between a split-phase motor and a capacitor-start motor?

5. What are some popular applications for the PSC motor?

6. How are PSC motors reversed? Can all PSC motors be reversed?

7. Explain how the speed of a PSC motor is changed.

8. Which winding in a split-phase motor has the largest resistance?

9. What is the purpose of a running capacitor in a PSC motor?

10. What is the speed of a six-pole, single-phase motor?

Three-Phase Electric Motors

Name: _____	Date: _____	Grade: ____

Comments:

Objectives: Upon completion of this lab, you should be able to correctly make the electrical connections to a three-phase motor and operate and reverse the motor.

Introduction: Three-phase motors are commonly used in the HVAC industry in larger systems such as light commercial, commercial, and industrial applications. Three-phase motors are considerably stronger and have more torque than single-phase motors because a three-phase electrical displacement is available because of the three legs of power being supplied to three-phase motors. Three-phase motors do not require starting components because of this displacement. Three-phase motors are rugged, reliable, and more dependable than other types of motors. Many three-phase motors utilized in the HVAC industry are dual voltage and can be operated on high or low voltage.

Text Reference: Paragraph 9.9

Tools and Materials: The following materials and equipment will be needed to complete this lab exercise.
Dual-voltage three-phase motor
Power supply to operate motor
Volt-ohmmeter
Clamp-on ammeter
Miscellaneous electrical handtools

Safety Precautions: Make certain that the electrical source is disconnected when making electrical connections. In addition:
- Make sure all connections are tight.
- Make sure no bare current-carrying conductors are touching metal surfaces.
- Make sure the correct voltage is being supplied to the motor.
- Make sure body parts do not come in contact with live electrical conductors.
- Keep hands and materials away from the rotating shaft.

LABORATORY SEQUENCE (mark each box upon completion of task)

A. Three-Phase Electric Motors

☐ 1. Obtain a dual-voltage three-phase motor from your instructor.

☐ 2. Check the resistance of the windings of the three-phase motor, record the resistance readings, and draw a wiring diagram of the motor.

Resistance readings _____

☐ 3. Make the proper electrical connections for the motor to operate on low voltage following the wiring diagram on the motor. If no diagram is available, refer to Figure 9.2 for diagrams of three-phase motors. The power source for this exercise is a disconnect supplied with 208/230-volt three-phase power.

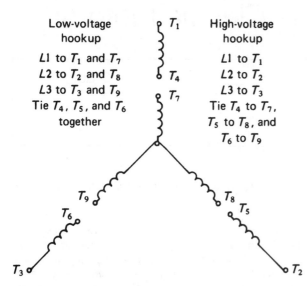

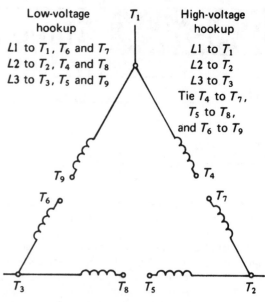

Low-voltage hookup

L1 to T_1 and T_7
L2 to T_2 and T_8
L3 to T_3 and T_9
Tie T_4, T_5, and T_6 together

High-voltage hookup

L1 to T_1
L2 to T_2
L3 to T_3
Tie T_4 to T_7,
T_5 to T_8, and
T_6 to T_9

Figure 9.2a Schematic diagram of the star winding of a three-phase motor.

Low-voltage hookup

L1 to T_1, T_6 and T_7
L2 to T_2, T_4 and T_8
L3 to T_3, T_5 and T_9

High-voltage hookup

L1 to T_1
L2 to T_2
L3 to T_3
Tie T_4 to T_7,
T_5 to T_8,
and T_6 to T_9

Figure 9.2b Schematic diagram of the delta winding of a three-phase motor.

Schematic of three-phase motor

☐ 4. Operate the motor, paying attention to the rotation. Record its current draw.

Current draw _____

☐ 5. Reverse the rotation of the motor by changing any two supply voltage conductors.

☐ 6. Operate the motor, paying attention to the rotation.

☐ 7. Remove the motor from the voltage source.

☐ 8. Make the electrical connections necessary for the motor to operate on high voltage.

☐ 9. Have your instructor check the wiring of the three-phase motor on high voltage.

☐ 10. Disconnect the connections made for high-voltage wiring and return the motor to the proper location.

☐ 11. Locate a three-phase motor on a unit in the lab and record its location.

_____

MAINTENANCE OF WORK STATION: Clean and return all tools to their proper location(s). Replace all equipment covers. Clean up the work area.

SUMMARY STATEMENT: Draw an electrical diagram of the winding of a dual-voltage three-phase motor. Explain how the motor can be used on low and high voltage by making adjustments to the electrical connections.

Questions

1. Why are no starting components needed on three-phase motors?

2. Explain the operation of a three-phase electric motor.

3. What is the limitation of using three-phase motors in the HVAC industry?

4. Explain how the same winding can be used on low and high voltage.

5. What are the two basic types of windings used in three-phase motors?

6. Give some applications of three-phase motors in the industry.

7. How can three-phase motors be reversed?

8. What are some advantages of using three-phase motors?

LAB 9–5 Electronically Commutated Motors

Name: _____	Date: _____	Grade: _____

Comments:

Objectives: Upon completion of this lab, you should be able to disassemble an ECM and examine the three parts (controller, rotor, and stator), change the controller, use an ohmmeter to determine if the motor section is good, and correctly install an ECM in an air-moving appliance. You will operate an HVAC system with an ECM and observe the operation and speed of the motor.

Introduction: The service technician will be required to install, replace, and troubleshoot ECMs in HVAC equipment. The ECM is a variable-speed motor and supplies a more comfortable environment and lower operation cost for the consumer. The controller and motor can be separated and the faulty section can be replaced without replacing both components. The technician must follow the equipment manufacturer's instructions when making the electrical connections.

Text References: Paragraph 9.10

Tools and Materials: The following materials and equipment will be needed to complete this lab exercise.
ECMs
Operating HVAC system with ECM
Volt-ohmmeter
Clamp-on ammeter
Miscellaneous electrical supplies
Miscellaneous electrical handtools
Velometer

Safety Precautions: Make certain that the electrical source is disconnected when making electrical connections. In addition:
- Make sure power is off before inserting or removing power connector on the ECM.
- Make sure all electrical connections are tight.
- Make sure no bare conductors are touching metal surfaces except the grounding conductor and the connection terminals.
- Make sure the correct voltage is being supplied to the equipment.
- Keep hands and other materials away from the rotating shaft.
- Make sure body parts do not come in contact with live electrical circuits.

LABORATORY SEQUENCE (mark each box upon completion of task)

A. Observing the Operation of an ECM

☐ 1. Obtain an operating HVAC system with an ECM from instructor.

☐ 2. Measure and record the air flow and wattage in Data Sheet 9F of the assigned unit at four different times, including startup.
   NOTE: The air volume (cfm) can be calculated by multiplying the velocity of air by the area of the opening in feet of the air flow outlet. The velocity of the air can be determined with a velometer.

After startup _____ cfm; _____ watts

10 minutes after startup _____ cfm; _____ watts

20 minutes after startup _____ cfm; _____ watts

30 minutes after startup _____ cfm; _____ watts

☐   3. Have your instructor check your data sheet.

B. Disassemble an ECM

☐   1. Obtain an ECM from your instructor.

☐   2. Examine the motor and locate the motor and motor control.

☐   3. Locate and remove the two ¼-inch hex head bolts at the back (opposite end of shaft) of the control housing. Make sure to prevent the motor or control from falling when the bolts are removed. NOTE: Remove only the hex head bolts and not the torx-head screws on older models.

☐   4. The control module is free from the mechanical attachment to the motor. Now carefully remove the plug from the motor to the control module. Do not pull on the wire grip, only the plug. The control module and motor are completely separated.

☐   5. Examine the outside of the control module for the control connector and power connector. The power connectors will have five terminals and the control connector will have sixteen terminals.

☐   6. Visually inspect the inside of the control module for signs of burned components or wires. If a component shows a sign of being overheated or burned, the control module is almost always defective.

☐   7. Check the resistance of the motor windings; they all should show equal resistance.

Resistance of motor windings 1 to 2 _____ ohms

Resistance of motor windings 1 to 3 _____ ohms

Resistance of motor windings 2 to 3 _____ ohms

Check to determine if the motor is grounded (and the electrical connection between the windings and motor housing).
NOTE: The resistance between the motor leads and the ground should be greater than 100,000.

Resistance of motor terminal 1 to housing _____ ohms

Resistance of motor terminal 2 to housing _____ ohms

Resistance of motor terminal 3 to housing _____ ohms

If the resistances of the motor windings are equal and measurable and the windings are not grounded, the motor is good.

☐   8. Mark the two end bells of the motor and the stator for correct placement when reassembly occurs.

☐ 9. Disassemble the motor by removing the remaining two ¼-inch hex head bolts. Examine the stator and rotor. Notice the three permanent magnets attached to the rotor by heavy-duty adhesive.

☐ 10. Assemble the motor, being careful not to overtighten the bolts holding the control module.

Questions

1. What is the resistance of each of the windings of an ECM?

2. How can a technician determine the power connection from the control connection on an ECM?

3. What type of motor is the ECM?

4. If you checked an ECM and the motor and bearing are good, what is the problem with the ECM?

5. What type of bearings are utilized in an ECM motor?

6. What precaution should a technician make when disconnecting the control module from the motor?

7. Draw a schematic diagram of the motor section of an ECM motor?

8. What is the advantage of using an ECM over a conventional PSC motor in an air-moving application?

LAB 9–6 Troubleshooting Electric Motors

Name: _____   Date: _____   Grade: _____

Comments:

Objectives: Upon completion of this lab, you should be able to correctly troubleshoot the common types of motors used in the HVAC industry.

Introduction: The service technician's responsibility is to diagnose and repair problems in HVAC equipment or control systems. Many times, these problems are motors that are not operating properly. The technician will have to determine if the electric motor is the problem or whether it is the control that is responsible for operating the motor. Once the determination has been made that the motor is faulty, the technician must replace the motor with a proper replacement.

Text References: Paragraphs 9.4 through 9.9

Tools and Materials: The following materials and equipment will be needed to complete this lab exercise.
Selection of loose motors including two each of the following motors: shaded pole, split phase, capacitor start, PSC, and three phase
Power supply to operate motors
Volt-ohmmeter
Clamp-on ammeter
Miscellaneous electrical handtools

Safety Precautions: Make certain that the electrical source is disconnected when making electrical connections. In addition:
- Make sure all connections are tight.
- Make sure no bare current-carrying conductors are touching metal surfaces except the grounding conductor.
- Make sure the correct voltage is supplied to the motor.
- Make sure body parts do not come in contact with live electrical conductors.

LABORATORY SEQUENCE (mark each box upon completion of task)

A. Troubleshooting Electric Motors

☐ 1. Obtain ten electric motors from your instructor.

☐ 2. Troubleshoot the ten motors by checking the condition of the bearings by physical inspection and the windings by making a resistance check.

☐ 3. Record the condition of each motor on Data Sheet 9G.

DATA SHEET 9G

Motor	Condition of Motor	Current Draw If Motor Is Good
#1	_____	_____
#2	_____	_____
#3	_____	_____
#4	_____	_____
#5	_____	_____

#6	_____	_____
#7	_____	_____
#8	_____	_____
#9	_____	_____
#10	_____	_____

☐   4. If the motor is okay, operate the motor and record the current draw on Data Sheet 9G.

☐   5. Have your instructor check your diagnosis of the electric motors.

☐   6. Return the ten electric motors to their proper location(s).

MAINTENANCE OF WORK STATION: Clean and return all tools to their proper location(s). Clean up the work area.

SUMMARY STATEMENT: Explain the proper procedure for troubleshooting electric motors.

Questions

1. How can a service technician determine the condition of the bearings in an electric motor?

2. How can a service technician determine the condition of the windings in an electric motor?

3. Explain the procedure a technician would use to check the resistance of a multispeed shaded-pole motor.

4. What external electrical components must be checked when troubleshooting a PSC or CS motor?

5. What method does a split-phase motor use to begin the initial rotation of the motor?

6. What electrical condition could exist when the windings of an electric motor are faulty?

7. How can a service technician check the centrifugal switch in a split-phase or CS motor?

8. When a service technician must replace an electric motor, what should be the primary concern?

9. What would be the resistance reading of a good winding in an electric motor?

10. What procedure should a service technician use to check the resistance of a three-phase motor?

LAB 9–7 Hermetic Compressor Motors

Name: _____   Date: _____   Grade: _____

Comments:

Objectives: Upon completion of this lab, you should be able to determine the C, S, and R terminals of a single-phase compressor, correctly wire a PSC hermetic compressor motor, and determine the condition of a hermetic compressor motor.

Introduction: Most compressor motors used in the HVAC industry are enclosed in a housing with the compressor connected by a direct-drive shaft. This type of compressor is called a *hermetic* or *semi-hermetic* compressor. The electrical connections of a hermetic motor are terminals that extend through the hermetic casing and are insulated with a glass material. Single-phase hermetic compressors usually have three terminals while three-phase hermetic compressors could have more because of a dual-voltage motor.

Text Reference: Paragraph 9.11

Tools and Materials: The following materials and equipment will be needed to complete this lab exercise.
    Hermetic compressors
    Power source to operate compressors
    Run capacitors for operating compressors
    Volt-ohmmeter
    Clamp-on ammeter
    Miscellaneous electrical supplies
    Miscellaneous electrical handtools

Safety Precautions: Make certain that the electrical source is disconnected when making electrical connections. Do not allow any of the oil from the hermetic compressor to come in contact with your skin. In addition:
- Make sure all connections are tight.
- Make sure no current-carrying conductors are touching metal surfaces except the grounding conductor.
- Make sure the correct voltage is being supplied to the unit.
- Make sure body parts do not come in contact with live electrical conductors.
- Keep hands and materials away from the rotating parts.

LABORATORY SEQUENCE (mark each box upon completion of task)

A. Determining the Common, Start, and Run Terminals of a Hermetic Compressor

☐   1. Obtain a selection of single-phase hermetic compressors from your instructor.

☐   2. If a compressor is installed in a unit, the wires must be removed from the compressor before ohm readings are taken.

☐   3. Read the resistance of the motor windings in the hermetic compressor. Make certain that good contact is made between the terminals of the compressor. (Figure 9.3) shows the internal windings of a single-phase hermetic compressor motor. (Figure 9.4) shows the resistive ohmic values of the single-phase hermetic compressor motor windings used for the following example.

     a. Find the largest reading between any two terminals. The remaining terminal is common [in Figure 9.4, the reading between A and B is largest (12 ohms); therefore, C is common].

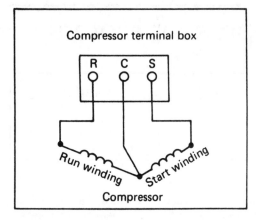

Legend

R: Run Terminal
C: Common Terminal
S: Start Terminal

Figure 9.3 Internal windings of a single-phase hermetic compressor motor.

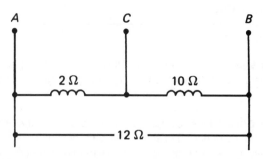

Figure 9.4 Terminals of a single-phase hermetic compressor motor with ohmic values given.

b. The largest reading between common and the other two terminals is start (C to A is 2 ohms and C to B is 10 ohms; therefore, common to B is larger, and B is start).

c. The remaining terminal is run. The reading between R and common is the smallest reading.

☐ 4. Measure and record the resistance of the terminals assigned to you by your instructor on Data Sheet 9H. Draw the terminal layout, labeling the resistances of the windings in the appropriate space.

DATA SHEET 9H

Compressor	Terminal Layout and Resistance
#1	_____
#2	_____
#3	_____
#4	_____
#5	_____

☐ 5. Determine the common, start, and run terminals of the hermetic compressor assigned and labelled in Data Sheet 9H.

☐ 6. Have your instructor check your terminal identification.

☐ 7. Return the compressors to their proper location, if necessary. Replace covers on any units where hermetic compressors were used.

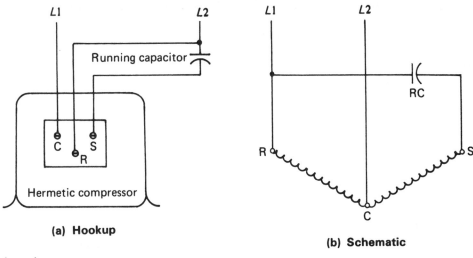

Legend

C: Common Terminal
R: Running Winding Terminal
S: Starting Winding Terminal
RC: Running Capacitor

Figure 9.5 (a) Hookup of a permanent split-capacitor motor (b) Schematic of a permanent split-capacitor motor.

B. PSC Hermetic Compressor Motors

☐ 1. Obtain the location of two single-phase hermetic compressors from your instructor.

☐ 2. Determine the common, start, and run terminals.

☐ 3. Make the proper electric connections in order to operate the single-phase hermetic compressors using the appropriate running capacitor. Figure 9.5 shows the diagram and wiring of a PSC hermetic compressor.

☐ 4. Connect each PSC hermetic compressor to the appropriate power supply.

☐ 5. Have your instructor check your wiring.

☐ 6. Operate each PSC hermetic compressor motor. Before operation, put a sock on the discharge line to prevent oil from spraying into the room. Record the current draw of each PSC hermetic compressor.

Current draw (Motor #1) _____

Current draw (Motor #2) _____

☐ 7. Return the hermetic compressors to their proper location, if necessary.

C. Troubleshooting Hermetic Compressors

☐ 1. Obtain the location of five hermetic compressors from your instructor.

☐ 2. Determine the condition of each compressor.

☐   3. Complete Data Sheet 9I with the diagnosis of the hermetic compressors.

DATA SHEET 9I

Compressor	Terminal Layout and Resistance
#1	_____
#2	_____
#3	_____
#4	_____
#5	_____

☐   4. Return the hermetic compressors to their proper location, if necessary.

MAINTENANCE OF WORK STATION AND TOOLS: Clean and return all tools and equipment to their proper location(s). Replace all equipment covers if any removed. Clean up the work area.

SUMMARY STATEMENT: What are the physical characteristics of a hermetic compressor motor? Why is it important for service technicians to be able to determine the common, start, and run terminals?

Questions

1. Why are all starting components except the motor winding external to a hermetic compressor motor?

2. Draw a wiring diagram of a PSC hermetic compressor with the power connections shown.

3. Find the common, start, and run terminals of the following hermetic compressors.

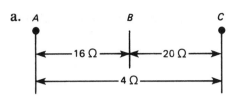

a.

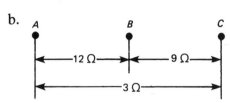

b.

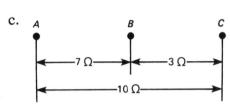

c.

4. Explain the procedure for troubleshooting hermetic compressor motors.

5. What are the four electrical failure categories for hermetic compressor motors?

6. What is a grounded hermetic compressor motor?

7. What is the highest allowable resistance reading for a grounded hermetic compressor motor?

8. What would the resistance readings of the windings be on a single-speed, three-phase hermetic compressor?

9. How would you diagnose a hermetic compressor with a faulty internal overload?

10. Why do some mechanical failures in hermetic compressors seem like electrical problems?

CHAPTER 10

Components for Electric Motors

Chapter Overview

Many single-phase electric motors used in hermetic compressors require some type of external starting component to control the start winding or other starting components used to assist the motor in starting. Many open-type electric motors use centrifugal switches that are mounted in the motor to accomplish the same task as these external components. Some open-type motors also use external components to control the start winding and starting components such as dishwasher motors, washing machine motors, drier motors, and other motors used where open contacts are not advisable. There are three types of starting relays used on single-phase enclosed motors: current (Figure 10.1), potential (Figure 10.2), and solid state (Figure 10.3). These devices are used on most single-phase hermetic compressor motors with the exception of permanent split-capacitor motors.

The method used by the starting relays to sense at what point in time the starting components need to be removed from the circuit is the major difference in starting relays. The start winding and starting components must be removed from the electrical circuit to prevent the motor from overloading, damaging the motor or destroying the starting capacitor. This is accomplished by dropping the start winding and starting components out of the electrical circuit once the motor reaches approximately 75% of full speed. The current or magnetic-type starting relay uses the current draw of the motor to energize and de-energize the start winding and starting components.

The potential or voltage relay uses the back electromotive force to energize a normally closed set of contacts to take the starting components out of the circuit. The solid-state relay uses a positive temperature coefficient (PTC) material that effectively removes the starting winding or component from the circuit. Each of these relays must be correctly sized and matched to the application, except the solid-state relays that can be used over certain horsepower ranges. Each relay is designed to remove the starting circuit when the motor reaches approximately 75% of full speed.

All rotating electric devices have some type of bearing to allow for smooth and easy rotation. The two types of bearings used in the HVAC industry are ball bearings and sleeve bearings. Ball bearings are the most efficient because they produce less friction and are used on heavy loads. The sleeve bearing is the most popular because of its low price and the fact that ball bearings could not be used in an enclosed application such as a hermetic compressor motor. Both types of bearings play an important part in motor, fan, and compressor operation in the industry. No matter what the type of bearing, proper lubrication is essential for long bearing life.

There are basically two types of motor drives used in the HVAC industry. A motor drive is the connection between an electric motor and a component that requires rotation. Electric motors are used to drive most devices that require rotation. There are two basic drive methods: direct

Figure 10.1 Current relay. *(Delmar/Cengage Learning)*

Figure 10.2 Potential relays. *(Delmar/Cengage Learning)*

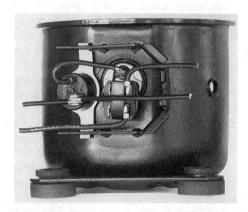

Figure 10.3 Solid-state relay installed on a compressor. *(Courtesy of Tecumseh Products Co.)*

drive and belt drive. The direct method couples the electric motor directly to the device requiring rotation. Some air-moving devices are attached directly to the shaft of the motor, while other devices requiring direct coupling use some type of coupling to connect the two devices together. Another popular method of driving devices that require rotation is the V-belt, which connects an electric motor with the device by a belt. The V-belt is widely used in the industry. Device speed can be changed by changing pulley sizes with the V-belt application. Device speed can only be changed by changing the motor speed on a direct-drive application.

Key Terms

Back electromotive force
Ball bearing
Current or magnetic relay

Direct drive
Potential relay
Sleeve bearing

Solid-state relay
Starting relays
V-belt

REVIEW TEST

Name: _____   Date: _____   Grade: ____

Fill in the blanks in questions 1 through 20 using the terms in following list.

168 volts
256 volts
420 volts
475
495 volts
5
525
6
625
7
75%
back electromotive force
ball bearings
belt tension
current relay
direct drive
lubrication
matched set
oil ring
oil wick
potential relay
proper alignment
PTC
sleeve bearings
V-belt
yarn packed

1. A solid-state starting relay uses a _____ material to disconnect the starting winding from a split-phase motor after starting.

2. Proper _____ is essential for the efficient operation and extended lifetime of electric motors.

3. Motor starting relays usually drop the starting components out of the circuit when the motor reaches approximately _____ of full speed.

4. A GE potential relay with the number 3ARR3-A6V5 has a coil with a continuous voltage rating of _____ volts.

5. A GE potential relay with the number 3ARR3-C2J6 has a coil with a continuous voltage rating of _____ volts.

6. An RBM potential relay with the number 128-126-2373BD has a coil with a continuous voltage rating of _____ volts.

7. An RBM potential relay with the number 128-112-1331JP has a coil with a continuous voltage rating of _____ volts.

8. The _____ is the voltage produced in the starting winding of a single-phase motor.

9. The _____ is a device that is designed to reduce friction in a rotating device and consists of balls enclosed by an inner and outer ring and lubricated with some type of lubricating grease.

10. The relay that uses the current produced by the motor to control the starting winding of a split-phase motor is a _____.

11. The resistance of the contacts of a potential relay should be _____ ohms.

12. When installing new V-belts on a dual-belt application, a _____ of belts should be used.

13. To insure long life of V-belts, the _____ _____ must be set properly.

14. The three methods used to lubricate sleeve bearings are _____, _____, and _____.

15. A _____ is used to transfer rotation between a motor with a pulley and a device equipped with a pulley.

16. A motor with a speed of 1750 rpm with a pulley diameter of 3 inches would rotate a fan with a pulley diameter of 10 inches at _____ rpm.

17. A rotating appliance requires a rotation of 1050 rpm if the motor is equipped with a 3-inch pulley. At 1750 rpm, a _____ inch pulley is required.

18. The back electromotive force operates the _____.

19. A fan motor connected to the shaft of a motor is considered a _____ application.

20. A _____ is an antifriction device that allows free turning and support of the rotating member. It consists of bronze or other low-friction material that is drilled to the diameter of the shaft.

Starting Relays for Single-Phase Electric Motors

Name: _____	Date: _____	Grade: _____
Comments:		

Objectives: Upon completion of this lab, you should be able to correctly install and troubleshoot a current, potential, and solid-state starting relay on a single-phase motor.

Introduction: Split-phase, capacitor-run, and capacitor-start–capacitor-run single-phase motors must have some type of electrical device that removes the starting winding and/or the starting capacitor from the circuit once the motor reaches approximately 75% of full speed. Current-type, potential, and solid-state relays are used to accomplish this in single-phase motors.

Text References: Paragraphs 10.1 through 10.5

Tools and Materials: The following materials and equipment will be needed to complete this lab exercise.
Selection of current relays including some good and bad relays
Selection of potential relays including some good and bad relays
Current, potential, and solid-state relays used to operate compressor motors
Power pack for PSC motor
Single-phase compressors (one with split-phase motor, one with capacitor-start motor, one with capacitor-start–capacitor-run motor)
Condensing unit with PSC hermetic compressor
Power source for compressor operation
Miscellaneous electrical supplies
Electrical meters
Miscellaneous electrical handtools

Safety Precautions: Make certain that the electrical source is disconnected when making electrical connections. In addition:
- Make sure all connections are tight.
- Make sure no bare ungrounded conductors are touching metal surfaces.
- Make sure the correct voltage is being supplied to the compressor.
- Make sure body parts do not come in contact with live electrical conductors.
- Keep hands and materials away from moving parts.

LABORATORY SEQUENCE (mark each box upon completion of task)

A. Troubleshooting Current-Type Relays

☐ 1. Obtain a selection of current-type and potential relays from your instructor.

☐ 2. Troubleshoot five current-type starting relays and record the condition of the relays in Data Sheet 10A. (NOTE: The current-type starting relay has a set of contacts and a coil. The coil can be checked with an ohmmeter. The resistance of the coil should be 0 ohms. The contacts of the current-type starting relay should read 0 ohms when the current-type relay is inverted. The current-type relay is normally open and the relay will have to be inverted to check the contacts.)

DATA SHEET 10A

Relay	Condition of Relay
#1	_____
#2	_____
#3	_____
#4	_____
#5	_____

B. Troubleshooting Potential Relays

☐ 1. Troubleshoot five potential-type starting relays and record the condition of the relays in Data Sheet 10B. (NOTE: The contacts of the potential relay are normally closed with the electrical connection from terminals 1 and 2. The coil of the potential relay should be checked for continuity with an ohmmeter. The resistance of some potential relay coil groups is very high. The electrical connection for the coil is from terminals 2 and 5.)

DATA SHEET 10B

Relay	Condition of Relay
#1	_____
#2	_____
#3	_____
#4	_____
#5	_____

☐ 2. Have your instructor check your diagnosis of the relays.

☐ 3. Clean up the work area and return all tools and supplies to their correct location(s).

C. Installing a Current-Type Relay on a Hermetic Compressor

☐ 1. Obtain from your instructor a fractional horsepower 115-volt hermetic compressor with a current relay that has external connections between the run and start terminals. These external connections allow for this type of current relay to drop the starting winding or both the starting winding and capacitor out of the starting circuit. If this type of relay is to be used with a split-phase motor, terminals 1 and 2 must be connected together with a conductor, as shown schematically in Figure 10.4. This type of relay can also be used for a capacitor-start motor by placing a capacitor between terminals 1 and 2, as shown in Figure 10.5.

☐ 2. Make the necessary electrical connections for the compressor to operate as a split-phase motor. Most current-type starting relays plug directly into the start and run terminals of the hermetic compressor. The common terminal of a single-phase hermetic compressor will be connected to one leg of power and in series with the overload on fractional horsepower compressors. If you have any problems in making the electrical connections, consult a wiring schematic of the compressor and relay. Terminals 1 and 2 of the current relay must be connected for the motor to operate as a split-phase motor.

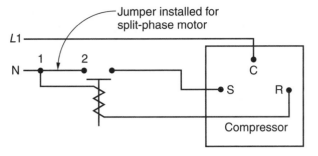

Figure 10.4 Current relay that can be used with a split-phase or CS motor wired as a split-phase motor.

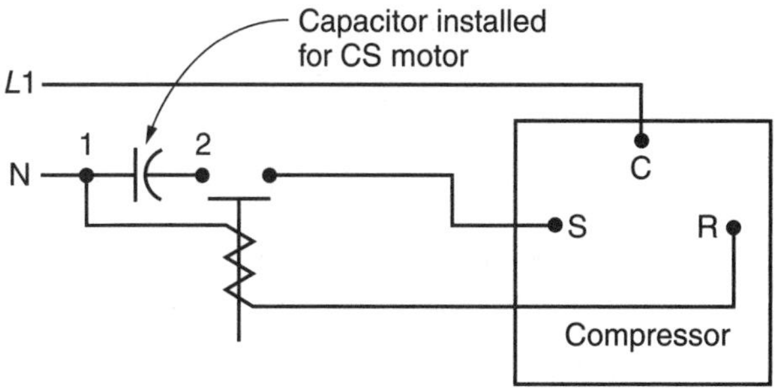

Figure 10.5 Current relay that can be used with a split-phase or CS motor wired as a CS motor.

3. Have your instructor check the wiring of the hermetic compressor and the starting relay before supplying power to the hermetic compressor.

4. After your instructor has checked the wiring of the compressor, operate the compressor and record the following information.

   Compressor model number _____

   Voltage and current draw of compressor _____

5. Remove the wiring from the hermetic motor.

6. Using the same fractional horsepower compressor, obtain the correct size starting capacitor according to the manufacturer's specifications.

7. Make the necessary electrical connections for the compressor to operate as a capacitor-start motor. The capacitor should be connected between terminals 1 and 2. The jumper on terminals 1 and 2 should be removed.

8. Have your instructor check the wiring of the hermetic compressor, current relay, and starting capacitor before supplying power to the compressor.

9. After your instructor has checked the wiring of the compressor, operate the compressor and record the following information.

   Compressor model number _____

   Voltage and current draw of compressor _____

10. Remove the wiring of the compressor and return all equipment to the proper location.

D. Installing a Potential Relay on a Hermetic Compressor

1. Obtain from your instructor a small 115-volt capacitor-start motor that requires a potential relay to drop the capacitor and starting winding from the start circuit. A schematic of a potential relay is shown in Figure 10.6.

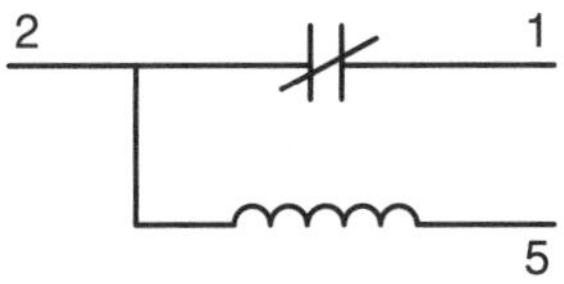

Figure 10.6 Schematic of potential relay.

2. Using the manufacturer's information, select the correct potential relay and starting capacitor for the compressor. Obtain a potential relay of the correct coil group and calibration number and capacitor for the compressor.

3. Correctly make the necessary connections for the compressor, potential relay, and capacitor.

4. Have your instructor check the wiring of the compressor and the starting components before supplying power to the compressor.

5. After your instructor has checked the wiring of the compressor, operate the compressor and record the following information.

   Compressor model number _____

   Potential relay coil group _____

   Voltage and current draw of compressor _____

6. Remove the wiring from the compressor and return all components to their correct location.

7. Obtain a 230-volt hermetic compressor from your instructor that can be operated as a PSC and CSR motor.

8. Using the manufacturer's information, select the correct potential relay and starting and running capacitors for your compressor. Many hermetic compressor motors can be wired as PSC motors and if the need arises for a high starting torque motor, a starting capacitor and potential relay can be added to the PSC motor to make it a CSR motor. A schematic of a hard-starting kit is shown in Figure 10.7.

9. Wire the compressor as a PSC motor using the correct running capacitor. Connect the power supply to the motor, but do not close the disconnect until the instructor has checked your electrical connections.

10. Operate the compressor and record the following information.

    Compressor model number _____

    Voltage and current draw of compressor _____

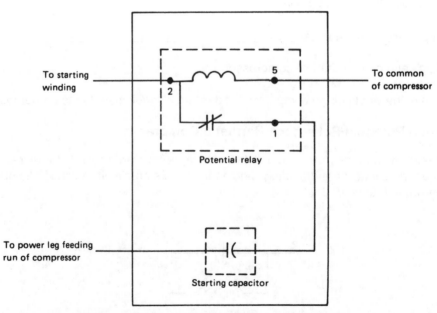

Figure 10.7 Schematic diagram of a hard-starting kit.

☐ 11. Add the necessary components to the PSC motor to convert the motor to a CSR motor. This will be the same procedure as adding a hard-start kit to the motor.

☐ 12. Have your instructor check your wiring.

☐ 13. Operate the compressor and record the following information.

Compressor model number _____

Potential relay coil group _____

Voltage and current draw of compressor _____

☐ 14. Remove the wiring and return all components to their proper location.

E. Installing a Solid-State Relay on a Hermetic Compressor

☐ 1. Obtain from your instructor a fractional horsepower split-phase compressor motor and a solid-state starting relay.

☐ 2. Make the necessary electrical connections following the wiring diagram and/or instructions that came with the relay.

☐ 3. Have your instructor check the wiring of the compressor and starting relay before supplying power to the hermetic compressor.

☐ 4. After your instructor has checked the wiring of the compressor, operate the compressor and record the following information.

Compressor model number _____

Relay model number _____

Voltage and current draw of compressor _____

☐ 5. Remove the wiring and return all components to their proper location.

☐ 6. Obtain a residential air-conditioning condensing unit assignment from your instructor. The unit should be operable and equipped with a PSC hermetic compressor motor.

☐ 7. Obtain a power pack from your instructor of the correct size for the assigned condensing unit.

☐ 8. Following the instructions included with the power pack, make the necessary electrical connections to install the power pack on the hermetic compressor motor.

☐ 9. Have your instructor check your wiring.

☐ 10. Start the condensing unit and record the following information.

Condensing unit model number _____

Model number of power pack _____

Voltage and current draw of condensing unit _____

☐ 11. Remove the power pack and return all components to their proper location.

☐ 12. Replace all covers on the equipment used for the lab project.

MAINTENANCE OF WORK STATION AND TOOLS: Clean and return all tools and supplies to their proper location(s). Replace all equipment covers. Clean up the work area.

SUMMARY STATEMENT: Why are relays needed to drop the starting winding and/or starting capacitor out of the starting circuit? What would be the results if a starting winding or starting capacitor remained in the circuit?

Questions

1. Explain the operation of a current-type starting relay.

2. On what applications are current-type starting relays likely to be found?

3. What is the advantage of using a PSC motor that can be converted to a CSR motor?

4. What is the proper procedure for troubleshooting a current-type starting relay?

5. What operates the coil on a potential relay?

6. Why are potential relay contacts normally closed and current-type starting relay contacts normally open?

7. Draw a wiring diagram of a CSR motor.

8. How can some current-type relays be used with split-phase and capacitor-start motors?

9. What is the advantage of using solid-state starting components?

10. Why is it necessary for some hermetic compressors to use high starting torque motors?

LAB 10–2 Electric Motor Bearings and Drives

Name: _____   Date: _____   Grade: _____

Comments:

Objectives: Upon completion of this lab, you should be able to identify, determine the condition, and lubricate the two types of motor bearings used in the industry. You will be able to correctly align a V-belt application and adjust a variable pitch pulley for a different speed without overloading the motor.

Introduction: All electric motors have some of type of bearings that reduce friction and allow for free rotation of the motor and other types of rotating devices. The two types of bearings used in the HVAC industry are sleeve bearings and ball bearings. Devices that require rotation must have some means of connecting the device to a source of rotating motion such as an electric motor or gasoline engine. The source of rotating power can be connected to the device by direct drive or V-belt; both methods are widely used in the industry.

Text References: Paragraphs 10.6 and 10.7

Tools and Materials: The following materials and equipment will be needed to complete this lab exercise.
    Electric motor with sleeve bearings
    Electric motor with ball bearings
    HVAC equipment with direct-drive and V-belt applications
    Miscellaneous hand tools for motor disassembly and pulley adjustments

Safety Precautions: Make sure that the electrical source is disconnected when examining live equipment. In addition:
   • Make sure all connections are tight.
   • Make sure no bare current-carrying conductors are touching metal surfaces except the grounding conductor.
   • Make sure the correct voltage is being supplied to the circuits.
   • Make sure no body parts come in contact with live electrical conductors.
   • Keep hands and materials away from moving parts.

LABORATORY SEQUENCE (mark each box upon completion of task)

A. Identifying Electric Motor Bearings

☐   1. Obtain two electric motors from your instructor, one with sleeve bearings and one with ball bearings.

☐   2. Mark the end bells and stators of the motors in order to determine their original location. This will be helpful during reassembly.

☐   3. Disassemble the electric motors, being careful to keep the parts of each motor separate.

☐   4. Identify the type of bearing and lubrication method in each type of motor and record the following information.

Motor	Type of Bearing	Lubrication Method
#1	_____	_____
#2	_____	_____

5. Reassemble the two motors, being careful to use the marks made in step 2. Once the motors have been reassembled, make certain that each turns freely.

6. Have your instructor check your identification and motor assembly.

7. Return the motors to their proper location.

B. Identification of Motor Drives

1. In the laboratory, locate at least five direct-drive applications and five V-belt applications.

2. Record the following information.

Unit	Type of Drive
#1	_____
#2	_____
#3	_____
#4	_____
#5	_____

3. Replace all equipment covers.

C. Adjusting a Variable-Pitch Pulley

1. Obtain a unit assignment from your instructor that has a variable pitch on a fan motor. The application must be a V-belt application.

2. Remove the cover from the fan compartment.

3. Locate the fan motor and fan. When the motor is started, visually observe the speed of the fan.

4. Operate and record the following information.

   Voltage of fan motor _____

   Current of fan motor _____

5. Turn the electrical power source off and tag the circuit.

6. Remove the V-belt from the fan pulley and motor pulley.

7. Inspect the variable-pitch pulley. The variable-pitch pulley has two Allen-head set screws. One set screw attaches the pulley to the motor shaft while the other set screw allows for one side of the pulley to be rotated. This rotation can make the pulley smaller or larger depending upon the direction of rotation. Locate the two set screws and determine which is responsible for the two operations.

8. Loosen the set screw holding the movable side of the pulley and turn the pulley counterclockwise until the pulley has been widened. This allows the belt to ride deeper in the pulley, thus making the pulley smaller and decreasing the revolutions per minute of the fan. When the fan motor is started, visually observe the speed of the fan.

☐ 9. Replace the V-belt on the fan and motor.

☐ 10. Restore electrical power to the unit.

☐ 11. Operate and record the following information.

  Voltage of fan motor _____

  Current of fan motor _____

☐ 12. Turn the electrical power off and tag the circuit.

☐ 13. Remove the V-belt from the fan and fan motor.

☐ 14. Loosen the set screw holding the movable side of the pulley and turn the pulley clockwise until the pulley has narrowed. This allows the belt to ride higher in the pulley, thus making the pulley diameter larger and increasing the revolutions per minute of the fan. When the fan motor is started, visually observe the speed of the fan.

☐ 15. Replace the V-belt on the fan and motor.

☐ 16. Restore electrical power to the unit.

☐ 17. Operate and record the following information.

  Voltage of fan motor _____

  Current of fan motor _____

☐ 18. Turn the electrical power off and tag the circuit.

☐ 19. Remove the V-belt from the fan and fan motor.

☐ 20. Return the pulley pitch to its original location.

☐ 21. Check the alignment of the pulley.

☐ 22. Correct alignment, if necessary.

☐ 23. Replace the V-belt on the fan and motor.

☐ 24. Check belt tension.

☐ 25. Correct belt tension, if necessary.

☐ 26. Restore electrical power to the unit.

☐ 27. Operate and record the following information.

  Voltage of fan motor _____

  Current of fan motor _____

Notice the amp draw of each pulley position. The higher the rpm, the higher the current draw.

☐ 28. Replace the covers on the fan compartment.

☐ 29. Have your instructor check your data.

☐ 30. Clean up the work area and return all tools and components to their proper location(s).

MAINTENANCE OF WORK STATION: Clean and return all tools to their proper location(s). Replace all equipment covers. Clean up the work area.

SUMMARY STATEMENT: Why is belt tension and alignment important? Give the advantages of direct-drive appliances and belt-drive appliances.

Questions

1. Explain how a variable-pitch pulley can change the revolutions per minute of a fan.

2. Why can't ball bearings be used in a hermetic compressor?

3. How are ball bearings lubricated?

4. How are sleeve bearings lubricated?

5. How can the rpm be changed in a direct-drive application?

6. Why are matched sets of V-belts used?

7. When are size A and B V-belts used?

8. What type of drive is used on most late model residential blower motors?

9. What type of drive is used to power a compressor on an automobile?

10. What type of coupling is used to connect an oil burner motor to the oil pump?

Chapter Overview

The control systems in modern heating, cooling, and refrigeration systems use many different control components to obtain automatic control of loads and thus safely maintain the temperature of a given space or substance. In many control systems, it is impractical to use a thermostat to directly control a device because of the current flow required. The larger the current draw a load produces, the larger the construction of contacts and mechanical linkage would have to be and the larger the components, the lower the accuracy of the thermostat. In many cases, air-conditioning control systems will use some type of pilot duty device such as a thermostat or pressure switch that will control a contactor or relay that can easily handle the current required by the electrical load. Smaller appliances such as domestic refrigerators, freezers, and window air conditioners utilize thermostats to directly control loads. Contactors and relays are primarily used in the industry to controls loads; these contactors and relays will be controlled by thermostats, pressure switches, and other switches.

A contactor is used to control an electric load by closing a set of contacts when voltage is applied to its coil or solenoid. A contactor consists of a coil that opens and closes a set of contacts due to the magnetic attraction created by the coil when the correct voltage is applied. Contactors are shown in (Figure 11.1). A schematic of a three-pole contactor is shown in (Figure 11.2).

The armature of a contactor is attached to the contacts so that when the armature is attracted to the magnetic field created when electrical energy is applied to the contactor coil, the contacts will close and when the coil is de-energized, a spring or gravity will open the contacts. The movement of the armature can be accomplished in two ways, using either a sliding or a swinging armature. Contactor coils are commonly available in 24, 115, 208, 230, 208/230, and 460 volts. In most cases, the coil is identified by the voltage marked on the coil. The contacts of the contactor are usually made of silver and cadmium, which resist sticking. The contacts are connected to a strong backing by mechanical or chemical bonding. Contactors are available with two, three, or four poles, which is the number of contacts. (Figure 11.3) shows a two-pole contactor controlling a compressor and condenser fan motor.

Relays are basically used for the same reason as the contactor to control a device by closing a set of contacts when voltage is applied to the relay coil. The major difference between a relay and contactor is the ampacity that they are capable of handling. Contactors in most cases are designed to carry larger loads than relays. The contact configuration of the relay is more complex than the contactor because of its function in control. Relays can be purchased with many different pole configurations. Normally open and normally closed contacts are both used in relays. The normally open contact closes when the coil is energized and opens when the coil is deenergized. The normally closed contact opens when the coil is energized and closes when the coil is deenergized. The normal position of relay contacts is in the deenergized position. The most common types of pole configuration for relays are single-pole–single-throw, single-pole–double-throw, double-pole–single-throw, and double-pole–double-throw. Several pole configurations are shown in (Figure 11.4). Relays are used to control the smaller loads in an HVAC control system.

Figure 11.1 Contactor.

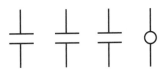

Figure 11.2 Schematic diagram of three-pole contactor.

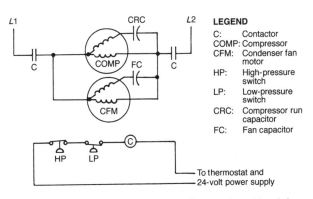

LEGEND

C: Contactor
COMP: Compressor
CFM: Condenser fan motor
HP: High-pressure switch
LP: Low-pressure switch
CRC: Compressor run capacitor
FC: Fan capacitor

Figure 11.3 Schematic diagram of a small residential air-cooled condensing unit with a contactor controlling the compressor and condenser fan motor.

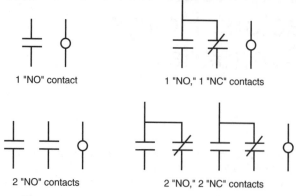

Figure 11.4 Common pole configuration of relays.

1 "NO" contact

1 "NO," 1 "NC" contacts

2 "NO" contacts

2 "NO," 2 "NC" contacts

An overload is an electrical device that protects a load from a high ampere draw by breaking a set of contacts. Overloads come in a variety of types, with the fuse being the simplest and some of the electronic overloads being the most complex. Fuses can be used to protect wires and noninductive loads such as electric heaters, but they provide inadequate protection for inductive loads such as motors. A load that is purely resistive such as an electric heater or light bulb with no coils to cause induction is called a *noninductive load*. Common resistive loads are incandescent light bulbs and resistance heaters. A load that uses coils of wire and creates magnetism is called an *inductive load*. Common inductive loads are motors, transformers, and solenoids. Inductive loads require overloads that will allow for the initial inrush current that is produced on the initial startup of an electric motor.

A line break overload opens on overload, removing the electrical energy from the circuit. The pilot duty overload breaks an auxiliary set of contacts that are in the control circuit. In most cases, a pilot duty overload opens the control circuit that is operating a contactor, which is in turn operating an electrical load.

A line break overload is sometimes called a *klixon*. This type of overload uses the current draw of a motor or load to heat a bimetal that will open a set of contacts, breaking the circuit going to the load. A line break overload installed on a hermetic compressor is shown in (Figure 11.5), and a schematic of this type of overload is shown in (Figure 11.6). Internal compressor overloads are a type of inline overload. A pilot duty overload has a set of contacts that open on an overload that is sensed by some type of element such as a thermal element or a magnetic element. A pilot duty overload is shown in (Figure 11.7) and its schematic is shown in (Figure 11.8). The electronic overload uses a sensor mounted in the motor winding that will sense the temperature of the motor windings and, during an unsafe condition, will open the control circuit. Most electronic overloads must have a power source if they are to be operable.

A magnetic starter is similar to a contactor with the exception of overload protection for the load. Many magnetic starters are controlled by push-button switches. (Figure 11.9) shows the schematic diagram of a push-button switch controlling a magnetic starter.

Figure 11.5 Inline overload of fractional horsepower compressor.

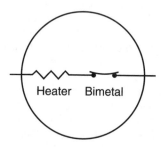

Heater   Bimetal

Figure 11.6 Schematic diagram of two-wire bimetal overload.

Figure 11.7 Current-type pilot duty overload.

Figure 11.8 Schematic diagram of a current-type pilot duty overload.

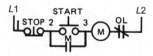

Figure 11.9 Schematic diagram of a push-button switch controlling a magnetic starter.

Electricity for Refrigeration, Heating, and Air Conditioning Lab Manual, Eighth Edition

Key Terms

Coil
Contactor
Contacts
Current overload
Fuses
Inductive load

Internal compressor overload
Line break overload
Magnetic overload
Magnetic starter
Mechanical linkage
Overload

Pilot duty overload
Push-button station
Relay
Resistive load

REVIEW TEST

Name: _____   Date: _____   Grade: ____

Answer the following questions.

1. What is the purpose of a contactor in an electrical control system?

2. Explain the operation of a relay.

3. What is the difference between a magnetic starter and a contactor?

4. What is the difference between a swinging and sliding armature when used in a contactor?

5. What is a resistive load?

6. What is an inductive load?

7. Draw a diagram of a contactor controlling a compressor using a line voltage thermostat.

8. What are the most common types of pole configurations used in relays?

9. How can the coil voltage of a relay or contactor be identified?

10. Draw a schematic diagram of a single-pole–double-throw relay controlling a blower motor, with the motor operating on high speed for cooling and low speed for heating.

11. What is the simplest form of overload protection used in the industry?

12. Why are single-element fuses used as overload protection with an electric resistance heater?

13. What is the difference between a line break overload and a pilot duty overload?

14. Why would a pilot duty overload be used instead of a line break overload?

15. What similarities exist between a line break bimetal overload and a pilot duty current-type overload?

16. What is the advantage of using an electronic overload on a large hermetic compressor?

17. How does an electronic overload sense that a hermetic motor is overloading?

18. Draw a diagram of a magnetic starter.

19. What types of overloads are used with magnetic starters?

20. What is the purpose of the auxiliary contacts when a magnetic starter is used with a push-button station?

LAB 11-1 Contactors and Relays

Name: _____   Date: _____   Grade: ___

Comments:

Objectives: Upon completion of this lab, you should be able to install contactors, relays, and overloads in HVAC equipment and draw their schematic diagrams.

Introduction: Contactors and relays are used in the HVAC industry to stop and start loads. The major difference between a contactor and a relay is the ampacity that the component can carry. Large loads such as compressors usually are controlled by contactors while small loads such as residential blower motors are controlled by relays. The technician must be able to install contactors and relays in HVAC equipment.

Text References: Paragraphs 11.1 and 11.2

Tools and Materials: The following materials and equipment will be needed to complete this lab exercise.
Relays and contactors with 24-, 115-, and 230-volt coils
24-volt control transformer
Volt-ohmmeter
Terminal boards
Switches
Basic electrical handtools
Cleat receptacles

Safety Precautions: Make certain that the electrical source is disconnected when making electrical connections. In addition:
- Make sure all connections are tight.
- Make sure no bare current-carrying conductors are touching metal surfaces except the grounding conductor.
- Make sure the correct voltage is being supplied to the circuits.
- Make sure body parts do not come in contact with live electrical conductors.
- Keep hands and materials away from moving parts.

LABORATORY SEQUENCE (mark each box upon completion of task)

A. 115-Volt Relay Connections

☐   1. Complete (Figure 11.10) so that the single-pole switch controls the relay that turns on the 115-volt light bulb. After you have made the proper connections on the pictorial diagram, draw the schematic.

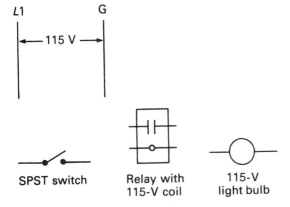

Figure 11.10

Schematic diagram

☐ 2. Have your instructor check both diagrams.

☐ 3. Using your diagram, wire the circuit on plywood board.

☐ 4. Connect the circuit to a 115-volt power source.

☐ 5. Have your instructor check your circuit.

☐ 6. Operate the circuit by closing the switch.

☐ 7. Record what happens when the switch is closed.

☐ 8. Disconnect the power source from the circuit and remove the components from the board.

B. 230-Volt Contactor Connections

☐ 1. Complete (Figure 11.11) so that the single-pole switch controls the contactor that turns on the 115-volt light bulb. After you have made the connections on the pictorial diagram, draw the schematic.

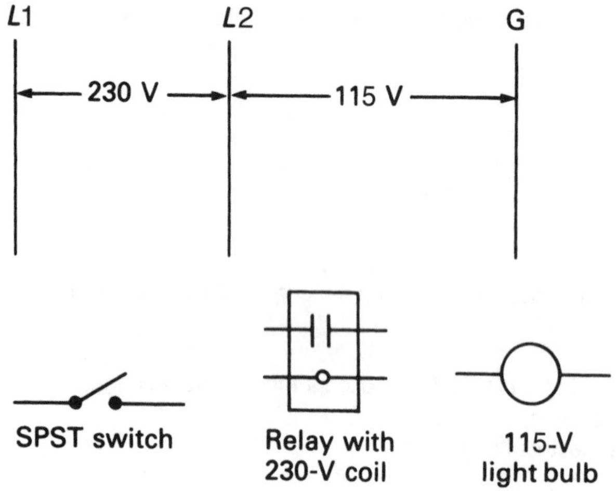

Figure 11.11

Schematic diagram

☐ 2. Have your instructor check both diagrams.

☐ 3. Using your diagram, wire the circuit on the board used in Part A, Step 3.

☐ 4. Connect the circuit to a 230-volt power source. (NOTE: The relay coil requires 230 volts, but the light bulb requires only 115 volts. The voltage from one leg of a 230-volt circuit to neutral is 115 volts.) Make your connections on the board.

☐ 5. Have your instructor check your circuit.

☐ 6. Operate the circuit by closing the switch.

☐ 7. Record what happens when the switch is closed.

☐ 8. Disconnect the power source from the circuit and remove the components from the board.

C. 24-Volt Relay Connections

☐ 1. Complete (Figure 11.12) so that the single-pole switch controls the relay that turns on the 115-volt light bulb. After you have made the connections on the pictorial diagram, draw the schematic.

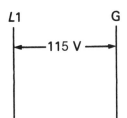

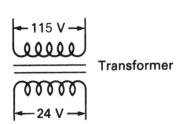

Transformer

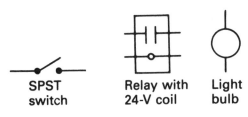

SPST switch   Relay with 24-V coil   Light bulb

Figure 11.12

Schematic diagram

☐ 2. Have your instructor check both diagrams.

☐ 3. Using your diagram, wire the circuit on the board used in Part A, Step 3.

☐ 4. Connect the circuit to a 115-volt power source. (NOTE: The relay coil requires 24 volts, which must be provided by the 24-volt transformer, but the light bulb requires 115 volts. The transformer must also be supplied with the correct power. Check the voltage to the transformer.) Make your connections on a terminal board.

☐ 5. Have your instructor check your circuits.

☐ 6. Operate the circuit by closing the switch.

☐ 7. Record what happens when the switch is closed.

☐ 8. Disconnect the power source from the circuit and remove the components from the board.

D. 24-Volt Relay, Using "NO" and "NC" Contacts of Relay

☐ 1. Complete (Figure 11.13) so that the single-pole switch controls the relay that energizes a 115-volt light bulb when the switch is open or closed. After you have made the connections on the pictorial diagram, draw the schematic.

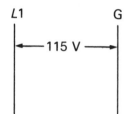

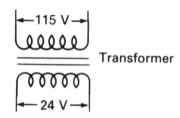

Transformer

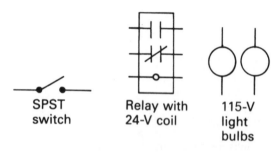

SPST
switch

Relay with
24-V coil

115-V
light
bulbs

Figure 11.13   Schematic diagram

☐ 2. Have your instructor check both diagrams.

☐ 3. Using your diagram, wire the circuit on the board used in Part A, Step 3.

☐ 4. Connect the circuit to a 115-volt power source. (NOTE: The circuitry is much the same as in Part C, Step 4 with the exception of the NC contacts.) Make your connections on a terminal board.

☐ 5. Have your instructor check your circuits.

☐ 6. Operate the circuit by closing the switch.

☐ 7. Record what happens when the switch is closed.

☐ 8. Disconnect the power source from the circuit and remove the components from the board.

MAINTENANCE OF WORK STATION AND TOOLS: Clean and return all tools to their proper location(s). Replace all equipment covers. Clean up the work area.

SUMMARY STATEMENT: What is the difference between a relay, contactor, and magnetic starter? Explain the application of each.

Questions

1. Why would a contactor be used on a three-phase load that has a full load amp draw of 65 amperes?

2. Why would a relay be used on a small residential blower motor?

3. What type of contactor would be used on a three-phase motor?

4. If information on a contactor stated 40A resistive load and 30A inductive load, could the contactor be used on a hermetic compressor that draws 40A?

5. How are the contacts of a contactor attached to their backing?

6. Of what are the contacts of a relay made?

7. What is the difference between a sliding armature and a swinging armature?

8. What type of relay would be used to control a blower motor on a residential furnace if the blower needs to operate on high speed during cooling operation and low speed during heating operation?

9. Draw a schematic diagram of a single-pole–double-throw relay.

10. Draw a schematic diagram of a four-pole contactor.

Name: _____	Date: _____	Grade: ____

Comments:

Objectives: Upon completion of this lab, you should be able to identify, understand, and install the types of overloads used in the HVAC industry.

Introduction: Most electrical loads are protected by an overload. Resistive loads are easy to protect in that they have a constant current draw, but inductive loads are harder to protect because their current draw is not constant and generally starts at a higher current, with the current diminishing as the load continues to operate. In most cases, resistive loads use fuses for protection and inductive loads use more complicated overloads, such as the bimetal, magnetic overload, current overload, and electronic. The service technician must be able to identify and replace the type of overload used. The technician must know what overload will the give the best protection in a particular application.

Text Reference: Paragraph 11.3

Tools and Materials: The following materials and equipment will be needed to complete this lab exercise.

Eight overloads for student identification   Clamp-on ammeter
Inline overloads   Volt-ohmmeter
Small hermetic compressor   Electrical handtools
Pilot duty current-type overloads   Miscellaneous wiring supplies
Small electric motor

Safety Precautions: Make certain that the electrical source is disconnected when making electrical connections. In addition:

- Make sure all connections are tight.
- Make sure no bare current-carrying conductors are touching metal surfaces except the grounding conductor.
- Make sure the correct voltage is being supplied to the circuits.
- Make sure body parts do not come in contact with live electrical conductors.
- Keep hands and materials away from moving parts.

LABORATORY SEQUENCE (mark each box upon completion of task)

A. Overload Identification

☐   1. Obtain eight overloads from your instructor.

☐   2. On Data Sheet 11A, record the type of each overload and whether it is inline or pilot duty.

DATA SHEET 11A

Overload	Type	Inline or Pilot Duty
#1	_____	_____
#2	_____	_____
#3	_____	_____
#4	_____	_____
#5	_____	_____
#6	_____	_____
#7	_____	_____
#8	_____	_____

☐ 3. Find and list the equipment location of each of the following overloads on Data Sheet 11B.

DATA SHEET 11B

Type of Overload	Equipment Model #
a. Inline	_____
b. Pilot Duty, Current-Type	_____
c. Electronic	_____

☐ 4. Have your instructor check your data sheet.

B. Inline Overloads

☐ 1. Draw a wiring diagram of an inline bimetal overload protecting a small hermetic compressor.

☐ 2. Have your instructor check your diagram.

☐ 3. Obtain an inline overload from your instructor and install it on a small hermetic compressor. (NOTE: You will have to install the correct starting relay on the compressor.)

☐ 4. Connect the hermetic compressor to a power source.

☐ 5. Have your instructor check your wiring.

☐ 6. Operate the motor and record the current draw.

Current draw of compressor _____

☐ 7. Disconnect the circuit from the power source.

☐ 8. Remove the components from the compressor and return them to their proper location.

C. Pilot Duty Overloads

☐ 1. Complete (Figure 11.14) by connecting the components to form a circuit to protect the motor. Draw a schematic of the circuit.

Electricity for Refrigeration, Heating, and Air Conditioning Lab Manual, Eighth Edition

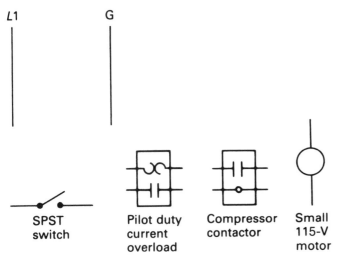

SPST switch · Pilot duty current overload · Compressor contactor · Small 115-V motor

Figure 11.14   Schematic diagram

☐ 2. Have your instructor check your diagrams.

☐ 3. Install the circuit on an electrical practice board.

☐ 4. Connect the circuit to a power source.

☐ 5. Have your instructor check your wiring.

☐ 6. Operate the motor and record the amp draw.

Current draw _____

☐ 7. Apply an artificial source of heat to the element of the current-type thermal overload until the pilot duty contacts open.

☐ 8. What happens to the electric motor?

☐ 9. Disconnect the circuit from the power supply.

☐ 10. Remove the components from the practice board and return them to their proper location.

MAINTENANCE OF WORK STATION AND TOOLS: Clean and return all tools to their proper location(s). Replace all equipment covers. Clean up the work area.

SUMMARY STATEMENT: Why are different overloads used for different applications?

Questions

1. Why are fuses used for protection of resistive loads?

2. What is the purpose of an overload?

3. What is a line break overload?

4. What is a pilot duty overload?

5. Why are pilot duty overloads used instead of line break overloads?

6. Explain the current draw of an electric motor from locked rotor to full speed.

7. What is an internal compressor overload?

8. What is the difference between a magnetic overload and a current-type pilot duty overload?

9. What is the advantage of using an electronic overload?

10. Draw a schematic diagram of a current-type pilot duty overload used to protect a hermetic compressor motor.

CHAPTER 12 | Thermostats, Pressure Switches, and Other Electric Control Devices

Chapter Overview

The function of most refrigeration, heating, or air-conditioning systems is to maintain the desired temperature of a space or object. The refrigeration, heating, and cooling industry uses many types of automatic controls to stop and start loads in equipment or systems and thus maintain the desired temperature and provide safe operation of major loads. The thermostat is the primary control used to control the temperature of structures, spaces, and objects. Pressure switches can also be used to control temperature by controlling pressure. Thermostats and pressure switches are also widely used as safety devices to prevent unsafe temperatures or pressures prevent in the equipment by stopping the affected load. Transformers are used in the industry to increase or decrease voltage to a desired level. Many other types of controls are used in the electrical systems on equipment such as humidistats, oil safety switches, time-delay relays, time clocks, and solenoid valves.

Thermostats play an important part in most control systems. The thermostat is commonly used as the primary temperature control. In some cases, it is also used as a safety device for motor protection, high temperature limits, and freeze protection. Thermostats that are primary controls can be heating thermostats or cooling thermostats or a combination of the two. Thermostats can also be used for staging equipment; that is, for operating parts of the equipment at different times, depending on the demands placed on the system.

Thermostats are designed to open and close a set of electrical contacts on a change in temperature. A heating thermostat closes on a decrease in temperature while a cooling thermostat closes on a rise in temperature, as shown in (Figure 12.1). Thermostats are available in many different types and styles. Thermostats use different types of elements to sense temperature, such as the bimetal element shown in (Figure 12.2) or the remote bulb. Both are widely used in the industry. Digital thermostats are becoming increasingly popular because of the decreased cost and their popularity. Various thermostats are shown in (Figure 12.3). Thermostats are available in line voltage, which is used in commercial applications, and low voltage, which is used in almost all residential applications and some commercial applications. Low-voltage thermostats are more accurate and perform more functions than line voltage thermostats.

(a) Heating thermostat; opens on temperature rise

(b) Cooling thermostat; closes on temperature rise

Figure 12.1 Symbols for heating and cooling thermostats.

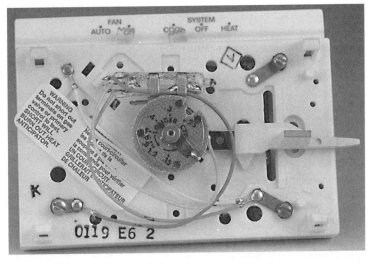

Figure 12.2 Thermostat with bimetal.

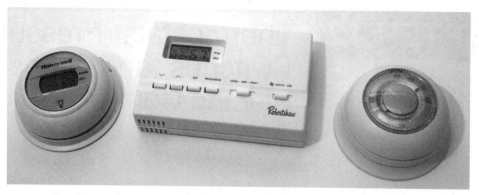

Figure 12.3 Thermostats.

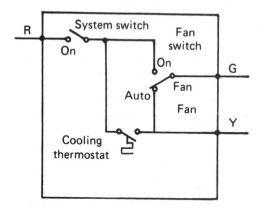

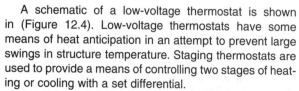

Figure 12.4 Schematic diagram of low-voltage thermostat.

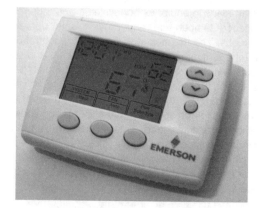

Figure 12.5 Programmable thermostat.

A schematic of a low-voltage thermostat is shown in (Figure 12.4). Low-voltage thermostats have some means of heat anticipation in an attempt to prevent large swings in structure temperature. Staging thermostats are used to provide a means of controlling two stages of heating or cooling with a set differential.

A programmable thermostat is used to control the temperature of a structure with the added convenience of allowing the customer to raise or lower the control point for one period or more in a 24-hour day, depending on the design and is shown in (Figure 12.5). The programmable thermostat gives the homeowner the flexibility of setting back the temperature in his or her home when the family is sleeping or away and returning the temperature to the desired level at a specific time.

Pressure switches are widely used in the industry. Pressure switches are designed to open and close on an increase or decrease in pressure, depending upon their application. A pressure switch is shown in (Figure 12.6). The symbols for pressure switches are shown in (Figure 12.7). Pressure switches come in a variety of pressure ranges; the technician must know the function of the pressure switch in the control system in order to determine the switching action and the pressure range. Pressure switches are sometimes used to control the temperature by using the pressure-temperature relationship and control other loads in the system. In most cases, pressure switches are used as safety controls to prevent damage to equipment when

Figure 12.6 Pressure switch.

the pressure is too high or low. The most important element of pressure switches is that the technician must know the application in order to understand their use.

Transformers decrease or increase the applied voltage to the desired voltage through the use of induction,

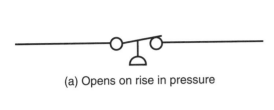

(a) Opens on rise in pressure

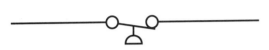

(a) Closes on rise in pressure

Figure 12.7 Symbols for pressure switches.

Figure 12.8 Transformer. *(Photo by bill Johnson)*

as shown in (Figure 12.8). Transformer ratings include primary voltage and frequency, secondary voltage, and voltamperes. Capacity is rated in voltamperes; that is, the voltage times the current of the control circuit loads. The more low-voltage loads in a control circuit, the larger the transformer that must be used.

Key Terms

Anticipators	Oil safety switch	Thermostat controlling
Clock thermostat	Pressure switch	element
Differential	Range	Time clock
Humidistat	Snap action	Time-delay relay
Line voltage thermostat	Solenoid valve	Transformer
Low-voltage thermostat	Staging thermostat	

REVIEW TEST

Name: _____   Date: _____   Grade: ____

Match the terms in the left-hand column to their definitions in the right-hand column.

_____ 1. Transformer

a. The pressure of the system when the pressure switch closes

_____ 2. Voltamperes (VA)

b. The temperature difference between the closing of the thermostat and the time when the warm air is no longer delivered to the room

_____ 3. Heating thermostat

c. Terminal on a thermostat representing the fan function

_____ 4. Cooling thermostat

d. An electrical device used to control the humidity in a structure

_____ 5. Remote bulb element

e. A device used to step-down voltage

_____ 6. Bimetal element

f. Opens on a temperature rise

_____ 7. Snap action

g. The element of a thermostat that senses the temperature through a liquid- and gas-filled bulb

_____ 8. System lag

_____ 9. Overshoot

_____ 10. Heating anticipator

_____ 11. Staging thermostat

_____ 12. Programmable
thermostat

_____ 13. "Y"

_____ 14. "W"

_____ 15. "G"

_____ 16. High-pressure
switch (safety)

_____ 17. Low-pressure
switch (safety)

_____ 18. Differential

_____ 19. Cut-in pressure

_____ 20. Humidistat

h. Opens on a rise in pressure

i. Transformer rating

j. Terminal on a thermostat representing the heating function

k. The temperature difference between the closing of the thermostat and the time when the warm air begins to reach the
thermostat

l. Direct closing of a thermostat without any floating of the
contacts

m. Closes on a rise in temperature

n. Adds artificial heat to the bimetal thermostat

o. Terminals on a thermostat representing the cooling function

p. Difference between the cut-in and cut-out settings

q. A thermostat that controls the temperature of a structure and
has the capability of setting the temperature back for a set
amount of time

r. A thermostat designed to operate more than one section of
equipment

s. Opens on a decrease in pressure

t. A thermostat element that uses two metals, each with a different expansion rate; this allows the element to respond to the
surrounding temperature

LAB 12–1   Transformers

Name: _____   Date: _____   Grade: ____

Comments:

Objectives: Upon completion of this lab, you should be able to install a control transformer in a circuit to energize a relay by closing a switch and install a transformer on a piece of equipment in the shop.

Introduction: Transformers are used primarily in the industry to supply a specific control voltage: 24 volts in residential and 115 volts in commercial and industrial equipment. Transformers are also used for many special applications, such as buck and boost, oil burner ignition, and electronic air filter power supplies. This lab will only cover the control transformer.

Text Reference: Paragraph 12.1

Tools and Materials:
24-volt secondary, 115-volt primary transformer
24-volt secondary, 208-volt primary transformer
24-volt secondary, 230-volt primary transformer
24-volt secondary, 115/208/230-volt primary transformer
115-volt secondary, 208/230-volt primary transformer

Relay with 24-volt coil
Relay with 115-volt coil
Volt-ohmmeter
Basic electrical hand tools

Safety Precautions: Make certain that the electrical source is disconnected when making electrical connections. In addition:
- Make sure all electrical connections are tight.
- Make sure no bare current-carrying conductors are touching metal surfaces except ungrounded conductors.
- Make sure the correct voltage is being supplied to the circuit or equipment.
- Make sure body parts do not come in contact with live electrical conductors.
- Keep hands and materials away from moving parts.

LABORATORY SEQUENCE (mark each box upon completion of task)

A. Ohmic Values of Transformers

☐ 1. Obtain the following transformers from your instructor:
24-volt secondary, 115-volt primary transformer
24-volt secondary, 208-volt or 230-volt primary transformer
24-volt secondary, 115/208/230-volt primary transformer
115-volt secondary, 208/230-volt primary

☐ 2. Using a volt-ohmmeter, take the resistance readings of the primaries and secondaries of the above transformers and record the readings in Data Sheet 12A. (NOTE: Do not touch the transformer leads when taking the resistance readings. Electrical shock can occur as the leads are disconnected.)

DATA SHEET 12A

Transformer	Primary Resistance	Secondary Resistance
24 V sec., 115 V prim.	_____	_____
24 V sec., 208/230 V prim.	_____	_____
24 V sec., 115/208/230 V prim.	_____	_____
115 V prim., 208/230 V sec.	_____	_____

☐   3. Have your instructor check your resistance readings.

☐   4. Review the resistance readings of the transformers. Technicians must understand completely what the resistance should be on a good transformer so they can troubleshoot transformers effectively.

☐   5. Keep the transformers used in this part of the lab for Part B.

B. Wiring a Transformer

☐   1. Using the transformers from Part A of this lab, connect the primary of each transformer to the correct voltage, the 24-volt secondary to a 24-volt relay coil, and the 115-volt secondary to a 115-volt relay coil. The relay solenoids should energize, closing the contacts of the relays. Have your instructor check the first relay before applying electrical energy. (Figure 12.6) shows an example of a 24-volt transformer connection to a 24-volt relay.

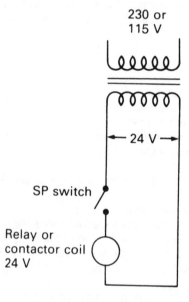

Figure 12.9

☐   2. Return all transformers to the proper location.

C. Install a Transformer in an Air-Conditioning Unit

☐   1. Obtain an air-conditioning unit from your instructor to replace the control transformer.

☐   2. Determine what type of transformer is needed for replacement.

☐   3. Disconnect the unit from the power source.

☐   4. Remove the old transformer from the unit, paying attention to the wiring connections.

☐   5. Install the new transformer.

☐   6. Have your instructor check your installation.

☐   7. Operate the unit, checking the current draw of all loads in the unit.

MAINTENANCE OF WORK STATION AND TOOLS: Clean and return all tools and supplies to their proper location(s). Replace all equipment covers. Clean up the work area.

Questions

1. What are some advantages of using a low-voltage control system?

2. What is the resistance of the secondary of a 24-volt transformer?

3. What is the resistance of a 115-, 208-, and 230-volt primary, 24-volt secondary transformer?

4. What is the purpose of using a fuse on the secondary side of the transformer?

5. What guidelines should be used when replacing a transformer in the field?

6. How are transformers rated?

7. What is the maximum load of a transformer used in an air-conditioning system?

8. What will happen if an undersized transformer is installed in a control system?

9. What size transformer is generally used on a heating-only gas furnace?

10. Draw a symbol for a transformer with a 24-volt secondary and a 115-, 208-, and 230-dual-voltage primary.

LAB 12–2　Low-Voltage Thermostats

Name: _____	Date: _____　Grade: ___

Comments:

Objectives: Upon completion of this lab, you should be able to correctly install a low-voltage thermostat and a programmable thermostat on a heating and cooling system and a heat pump in the lab.

Introduction: Low-voltage thermostats are one of the most used electrical devices in the heating, cooling, and refrigeration industry. Low-voltage thermostats serve many functions in equipment operation in modern residential and light commercial heating and cooling system applications, such as the control of heating, cooling, and fan functions. Almost all residences use a low-voltage thermostat because of more accurate control and other advantages. The technician must be able to correctly install low-voltage thermostats.

Text References: Paragraphs 12.2, 12.3, and 12.4

Tools and Materials: The following materials and equipment will be needed to complete this lab exercise.
    Single-stage heating and cooling low-voltage thermostats
    Two-stage heating and two-stage cooling thermostats
    Electronic programmable thermostat
    Other electrical components as determined by the diagrams
    Air-conditioning system using single-stage heating and cooling thermostat
    Heat pump
    Volt-ohmmeter
    Miscellaneous wiring supplies
    Basic electric hand tools

Safety Precautions: Make certain that the electrical source is disconnected when making electrical connections. In addition:
- Make sure all electrical connections are tight.
- Make sure no bare current-carrying conductors are touching metal surfaces except ungrounded conductors.
- Make sure the correct voltage is being supplied to the circuit or equipment.
- Make sure body parts do not come in contact with live electrical conductors.
- Keep hands and materials away from moving parts.

LABORATORY SEQUENCE (mark each box upon completion of task)

A. Single-Stage Heating and Cooling, Low-Voltage Thermostat Circuitry

☐　1. Complete (Figure 12.7) (next page) to form an electrical control system to control the heating, cooling, and fan. Light bulbs represent system loads, such as compressors, fan motors, and heating sources. After you have completed the connections on the pictorial diagram, draw a schematic of the control system circuitry.

☐　2. Have your instructor check both diagrams.

☐　3. Using your diagram, wire the control system on an electrical practice board. Read the installation instructions provided with the thermostat.

☐　4. Connect your system to a 115-volt power supply.

☐　5. Have your instructor check your circuit.

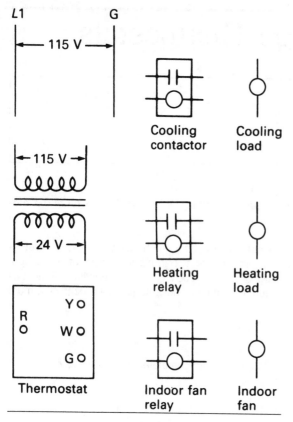

L1   G

|←— 115 V —→|

Cooling contactor   Cooling load

|← 115 V →|

|← 24 V →|

Heating relay   Heating load

Y ○
R ○   W ○
G ○

Thermostat

Indoor fan relay   Indoor fan

Figure 12.10   Schematic diagram

☐   6. Operate the control system by setting the thermostat for the desired function; check cooling, heating, and fan operation.

☐   7. Record what happens when the thermostat is set for each function.

☐   8. Disconnect the power source from the circuit and remove the components from the board.

B. Single-Stage Heating and Cooling, Low-Voltage Thermostat Installation

☐   1. Obtain an equipment assignment and thermostat from your instructor for installation of a single-stage heating and cooling thermostat.

☐   2. Read the installation instructions for the assigned equipment and thermostat.

☐   3. Disconnect the electrical power source from each piece of equipment.

☐   4. Remove the necessary covers from the equipment.

☐   5. Make the low-voltage connections between the equipment and the thermostat for proper operation.

☐   6. Have your instructor check your wiring.

☐   7. Restore electrical power to the equipment.

8. If the fan compartment cover was removed from a fossil fuel furnace, this cover must be replaced before operating the system.

9. Operate the system by setting the thermostat for the desired function; check cooling, heating, and fan operation.

10. Write a brief paragraph explaining the operation of the system.

11. Replace all covers on equipment.

12. Disconnect the electrical power supply from the equipment.

C. Two-Stage Heating and Cooling, Low-Voltage Thermostat Installation

1. Complete (Figure 12.8) (next page) to form an electrical control system to control the heating, cooling, and fan of a two-stage heating, two-stage cooling, low-voltage thermostat. Light bulbs represent system loads, such as compressors, fan motors, and heating sources. After you have completed the pictorial diagram, draw a schematic of the control system circuitry.

2. Have your instructor check both diagrams.

3. Using your diagram, wire the control system on an electrical practice board. Read the installation instructions provided with the thermostat.

4. Connect your system to a 115-volt power supply.

5. Have your instructor check your circuits.

6. Operate the control system by setting the thermostat for the desired function; check cooling, heating, and fan operation.

7. Record what happens when the thermostat is set to each function. Note the staging of the thermostat.

8. Disconnect the power supply and remove the components from the board.

D. Two-Stage Heating, Single-Stage Cooling, Low-Voltage Thermostat Installation

1. Obtain equipment assignment and thermostat from your instructor for installation of two-stage heating, single-stage cooling thermostat on a heat pump.

2. Read installation instructions for assigned equipment and thermostat.

3. Disconnect electrical power source from each piece of equipment.

4. Remove necessary covers from equipment.

5. Make the necessary low-voltage connections between the equipment and the thermostat for proper operation.

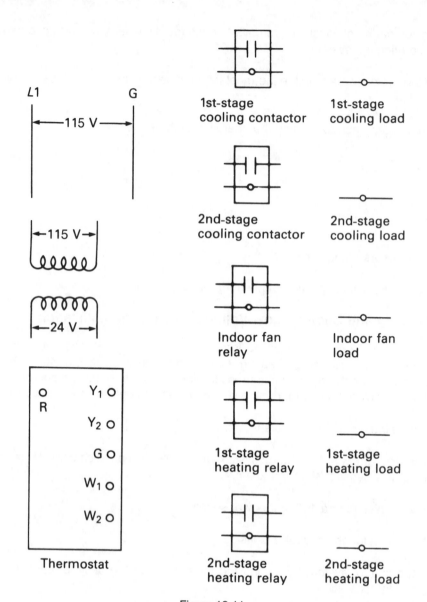

L1   G

|←——115 V——→|

|←—115 V—→|

|←—24 V—→|

| Thermostat |
O R	Y₁ O
	Y₂ O
	G O
	W₁ O
	W₂ O

1st-stage
cooling contactor

1st-stage
cooling load

2nd-stage
cooling contactor

2nd-stage
cooling load

Indoor fan
relay

Indoor fan
load

1st-stage
heating relay

1st-stage
heating load

2nd-stage
heating relay

2nd-stage
heating load

Figure 12.11

Schematic Diagram

Electricity for Refrigeration, Heating, and Air Conditioning Lab Manual, Eighth Edition

☐   6. Have your instructor check your wiring.

☐   7. Restore electrical power to the equipment.

☐   8. Replace indoor unit covers.

☐   9. Operate the system by setting the thermostat for the desired function: cooling, heating, second-stage heat, and fan operation.

☐   10. Write a brief paragraph explaining the operation of the system.

☐   11. Replace all equipment covers.

☐   12. Disconnect the electrical power supply from the equipment.

E. Installing an Electronic Programmable Thermostat

☐   1. Obtain an equipment assignment and thermostat from your instructor for installation of an electronic programmable thermostat on a single-stage heating and cooling system.

☐   2. Read the installation instructions for the assigned equipment and thermostat.

☐   3. Disconnect the electrical power source from each piece of equipment.

☐   4. Remove the necessary covers from the equipment.

☐   5. Make the necessary low-voltage connections between the equipment and the thermostat for proper operation.

☐   6. Have your instructor check your wiring.

☐   7. Restore electrical power to the equipment.

☐   8. If the fan compartment cover was removed from a fossil fuel furnace, this cover must be replaced before operating the system.

☐   9. Review the instructions on setting the thermostat.

☐   10. Program the thermostat as assigned by the instructor.

☐   11. Operate the system by setting the thermostat for the desired function; check cooling, heating, and fan operation.

☐   12. Write a brief paragraph explaining the operation of the system.

☐   13. Replace all covers on equipment.

☐   14. Disconnect the electrical power supply from the equipment.

MAINTENANCE OF WORK STATION: Clean and return all tools to their proper location(s). Replace all equipment covers. Clean up the work area.

SUMMARY STATEMENT: Why are low-voltage thermostats used for the control of most residential heating and cooling systems?

Questions

1. How do heating and cooling thermostats differ?

2. What is the purpose of a heat anticipator in a thermostat?

3. Why is it important that thermostat contacts are snap acting?

4. Draw a simple diagram of a single-stage heating and cooling thermostat.

5. What is the purpose of the letter designations on low-voltage thermostats?

6. Briefly explain proper installation procedures for installing thermostats.

7. What is a staging thermostat?

8. Why are staging thermostats used on heat pumps?

9. What is the purpose of using an electronic programmable thermostat?

10. Explain the operation of an electronic programmable thermostat.

LAB 12–3 Line Voltage Thermostats

Name:	Date:	Grade:

Comments:

Objectives: Upon completion of this lab, you should able to correctly install a line voltage thermostat on a unit in the shop.

Introduction: Line voltage thermostats are widely used in the refrigeration, heating, and air-conditioning industry. They are used as operating controls on domestic refrigerators and freezers, in commercial refrigeration appliances to maintain the temperature in walk-in-coolers and walk-in-freezers, and in other applications where temperature is the element that must be controlled such as limit switches and fan switches on fossil fuel furnaces. The technician must know the function of the line voltage thermostat in the control system.

Text Reference: Paragraph 12.2

Tools and Materials:
Line voltage thermostats
Other components as determined by diagrams
Volt-ohmmeter

Miscellaneous wiring supplies
Basic electrical hand tools

Safety Precautions: Make certain that the electrical power source is disconnected when making electrical connections. In addition:
- Make sure all connections are tight.
- Make sure no bare current-carrying conductors are touching metal surfaces except the grounding conductor.
- Make sure body parts do not come in contact with live electrical conductors.
- Keep hands and materials away from moving parts.

LABORATORY SEQUENCE (mark each box upon completion of task)

A. Line Voltage Thermostat Circuitry

☐ 1. Complete Figure 12.12 so that the line voltage thermostat energizes a small 115-volt fan motor when the temperature exceeds 85°F. After you have made the proper connections on the pictorial diagram, draw the schematic.

☐ 2. Have your instructor check both diagrams.

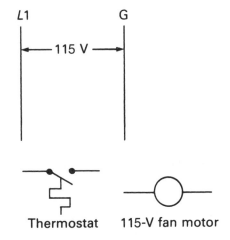

Figure 12.12

Schematic diagram

☐ 3. Using your diagrams, wire the circuit on an electrical practice board.

☐ 4. Connect the circuit to a 115-volt power source.

☐ 5. Have your instructor check your circuit.

☐ 6. Operate the circuit by closing the thermostat.

☐ 7. Record what happens when the thermostat closes.

☐ 8. Disconnect the power source from the circuit and remove the components from the board.

B. Line Voltage Thermostat Installation

☐ 1. Obtain an equipment assignment and line voltage thermostat from your instructor for installing a line voltage thermostat to control a load.

☐ 2. Draw a schematic diagram of the application your instructor has assigned.

☐ 3. Have your instructor check your wiring diagram.

☐ 4. Disconnect the power supply from the equipment.

☐ 5. Install a line voltage thermostat to meet the specifications of your instructor.

☐ 6. Have your instructor check your wiring.

☐ 7. Operate the equipment and check the thermostat.

☐ 8. Write a brief paragraph explaining the function of the line voltage thermostat in the circuit.

☐ 9. Disconnect the equipment from the power source.

MAINTENANCE OF WORK STATION: Clean and return all tools to their proper location(s). Replace all equipment covers. Clean up the work area.

SUMMARY STATEMENT: Give three applications each of line voltage thermostats in residential air-conditioning and heating systems and commercial refrigeration systems.

Questions

1. Why are line voltage thermostats used to control domestic refrigerators?

2. What are some disadvantages of using a line voltage thermostat?

3. Draw a schematic of a line voltage thermostat starting a condenser fan motor when the temperature reaches 90°F.

4. What is the advantage of using a line voltage thermostat over a low-voltage thermostat?

5. Why are line voltage thermostats less accurate than low-voltage thermostats?

6. Draw a schematic of a line voltage thermostat starting a blower motor when the furnace combustion chamber gets warm.

7. Draw a schematic of a line voltage thermostat interrupting the power supply to a line voltage gas valve if the furnace combustion chamber overheated.

8. Why are line voltage thermostats widely used in commercial refrigeration equipment?

9. What is the difference between the bimetal used for a low-voltage thermostat and a line voltage thermostat?

10. What are some common controlling elements used with line voltage thermostats?

LAB 12–4 Pressure Switches

Name: _____	Date: _____	Grade: ____

Comments:

Objectives: Upon completion of this lab, you should be able to correctly adjust and install a pressure switch to function as an operating and safety control.

Introduction: Pressure switches are used in control circuits as safety and operating controls. When a pressure switch is used as a safety control, it de-energizes a control circuit, interrupting the power source to a major load when an unsafe condition exists. An operating control starts and stops a load for a particular reason, such as head pressure control or temperature control. It is important that the technician knows the function of the pressure switch in the control circuit.

Text Reference: Paragraph 12.5

Tools and Materials: The following materials and equipment will be needed to complete this lab exercise.

Eight pressure switches for identification
High-pressure switch (opens on rise)
Low-pressure switch (closes on rise)
Dual-pressure switch

Operating systems for installation of pressure switches
Volt-ohmmeter
Miscellaneous wiring supplies
Basic electrical hand tools

Safety Precautions: Make certain that the electrical source is disconnected when making electrical connections. In addition:

- Make sure all electrical connections are tight.
- Make sure no bare current-carrying conductors are touching metal surfaces except ungrounded conductors.
- Make sure the correct voltage is being supplied to the circuit or equipment.
- Make sure body parts do not come in contact with live electrical conductors.
- Keep hands and materials away from moving parts.

LABORATORY SEQUENCE (mark each box upon completion of task)

A. Identifying Pressure Switches

☐   1. Obtain eight pressure switches from your instructor.

☐   2. Record the type, function, and setting of each switch on Data Sheet 12B.

DATA SHEET 12B

Switch	Function	Setting
#1	_____	_____
#2	_____	_____
#3	_____	_____
#4	_____	_____
#5	_____	_____
#6	_____	_____
#7	_____	_____
#8	_____	_____

3. Find and list the equipment location of each of the following pressure switches in the lab on Data Sheet 12C.

DATA SHEET 12C

Adjustable high-pressure switch (safety) _____

Fixed high-pressure switch (safety) _____

Adjustable high-pressure switch (operating) _____

Adjustable low-pressure switch (safety) _____

Fixed low-pressure switch (safety) _____

Adjustable low-pressure switch (operating) _____

Dual pressure switch (safety) _____

4. Have your instructor check your data sheet.

B. Low-Pressure Switch Used as an Operating Control

1. Obtain instructor assignment of a commercial refrigeration system to set a low-pressure switch to maintain the temperature of a medium-temperature application to instructor specifications.

2. Set the pressure switch to obtain the assigned temperature.

3. Operate the unit and record the temperature and off cycle time.

   Temperature _____ Off cycle time _____

4. Have your instructor check your work.

C. High-Pressure Switch Used as an Operating Control

1. Obtain from your instructor a unit assignment and a high-pressure switch that closes when the temperature rises.

2. Disconnect the power supply from the equipment.

3. Install the high-pressure switch and make all necessary connections (pressure and electrical). Make certain that you have closed all necessary valves and isolated the section of the refrigeration system where the switch is to be installed.

4. Set the pressure switch for the assigned pressure.

5. Have your instructor check your pressure connections, wiring connections, and pressure switch setting.

6. Connect the power supply to the system.

7. Operate the system, observe the actions of the pressure switch, and record the pressures when the condenser fan cuts off and on.

   Fan off _____ Fan on _____

8. Disconnect the power supply from the equipment.

D. Dual-Pressure Switch Used as a Safety Control

☐   1. Obtain a dual-pressure switch and a unit assignment from your instructor.

☐   2. Disconnect the equipment from the power supply.

☐   3. Install the dual-pressure switch and make all necessary connections (pressure and electrical). Use valves to isolate parts of the refrigeration system in order to make pressure connections.

☐   4. Set and record the pressure switch for the proper safety setting.

      High-pressure setting _____

      Low-pressure setting _____

☐   5. Have your instructor check your pressure connections, wiring connections, and setting of the pressure switch.

☐   6. Connect the power supply to the equipment.

☐   7. Operate the system and observe the actions of the pressure switch.

☐   8. Block the condenser with a piece of cardboard and operate the system. Record the pressure at which the equipment cuts off. If the discharge pressure exceeds safe limits as set by the instructor, cut the unit off using the disconnect.

      Cut-out pressure _____

☐   9. Remove the cardboard and return the system to normal operation.

☐   10. Close the compressor suction service valve and operate the system. Record the pressure at which the equipment cuts off. If the suction pressure goes below safe limits as set by the instructor, cut the unit off using the disconnect.

      Cut-out pressure _____

☐   11. Open the suction service valve and return the system to normal operation.

☐   12. Disconnect the equipment from the power supply.

MAINTENANCE OF WORK STATION AND TOOLS: Clean and return all tools to their proper location(s). Replace all equipment covers. Clean up the work area.

SUMMARY STATEMENT: Why does the pressure setting change on pressure switches when they are used as safety controls with different refrigerants?

Questions

1. What would be the action of a low-pressure switch used as an operating control?

2. What would be the low-pressure cut-out for a low-pressure switch used as a safety control?

3. What is the difference between fixed and adjustable high-pressure switches?

4. What factors must the technician know about a pressure switch before obtaining a replacement?

5. What would be the cut-out setting of a high-pressure switch used on an R-22 system?

6. What is the differential of a pressure switch?

7. What would be the cut-in and cut-out settings of a low-pressure switch used as a safety control on an R-22 system?

8. In what applications are high-pressure switches used as operating controls?

9. In what applications are low-pressure switches used as operating controls?

10. Why is it important for the technician to know the function of a pressure switch in a control circuit?

Chapter Overview

The refrigeration, heating, and air-conditioning industry has moved rapidly to incorporate electronic control circuits into many of the control systems being used today. Improvements and new innovations made in electronic components have revolutionized control systems and components in the industry, yielding smaller, more accurate, and more diversified electronic components. Modern electronic control systems produce better control parameters, more efficient operation, and multifunctional control systems. Along with better control, many other advances have been made: in timed controls circuits, in motor speed control, in motor protection, in defrost systems for heat pumps and commercial refrigeration, in ignition controls for gas furnaces, and others. Most air-conditioning manufacturers use electronic control boards to supervise the operation of heat pumps, gas furnaces, and zoned systems.

Major improvements have been made in commercial and industrial equipment by using electronic control systems. In the past, controls for these units were large and bulky and often did not provide adequate control of each zone of the structure. Current electronic control systems and components have almost eliminated erratic control of individual zones in commercial and industrial applications. Many of these units also incorporate a method of controlling the fan speed, depending on the load of the building and zones by using electronic speed controls.

Residential heating and air-conditioning equipment and control systems have also been affected by the wide use of electronic control modules. At present, many residential heating and cooling systems use electronic controls to supervise the operation of a gas furnace as shown in (Figure 13.1), to supervise the operation of heat pumps as shown in (Figure 13.2), to control the operation of an oil burner as shown in (Figure 13.3), to control the speed of variable-speed motors in air conditioners and heat pumps, and in many other applications. Microprocessors are currently being used to control the operation of equipment and entire control systems. Many control manufacturers market an electronic zone control system that is used in many residential installations. The use of solid-state controls in residential systems has increased efficiency and produced better control for the customer.

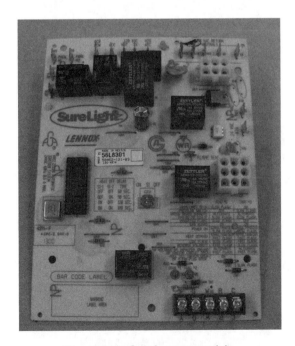

Figure 13.1 Gas furnace module.

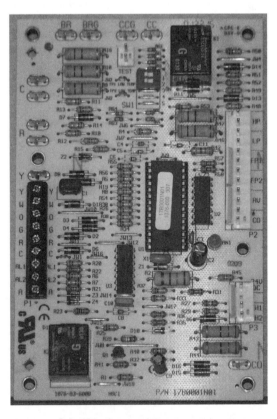

Figure 13.2 Heat pump module.

Figure 13.3 Oil burner control.

Figure 13.5 Electronic time-delay relay.

The heating and air-conditioning industry has developed many single-function solid-state modules that are in common use today. Single-function solid-state modules are being used in almost all phases of the industry. A single-function defrost control used on a heat pump is shown in (Figure 13.4). Electronic time-delay devices are widely used in the industry as anti-short-cycling devices, time delay relays, and in other devices requiring a time function. An electronic time delay relay is shown in (Figure 13.5). There are many types of electric motor protection devices that are being used in the industry today to protect electric motors against overload. An electronic motor protection module is shown in (Figure 13.6). Solid-state defrost modules are commonly used on heat pumps and in commercial refrigeration units, where defrosting of a coil is essential to the efficient operation of the equipment. Heating equipment uses many types of electronic ignition modules that supervise the operation of gas and oil burners. An ignition module used on a gas furnace is shown in (Figure 13.7).

The commercial refrigeration sector of the industry is beginning to utilize solid-state control systems. Almost all control systems used in the transport refrigeration industry are solid state because of the adaptability with

Figure 13.6 Electronic motor protector.

Figure 13.4 Single-function defrost control.

Figure 13.7 Ignition module.

the direct current that is used. Many specialized single-function solid-state devices are used in the stationary commercial refrigeration market, such as oil pressure controls, motor protection modules, low-pressure switch modules, and time-delay modules. Electronic control systems have become extremely popular in large structures using variable air volume systems. As technology advances, we will see even further development of more and better electronic control systems.

These new solid-state control systems are not unlike conventional controls in that they require servicing. Although the terms "solid-state" or "electronic" may intimidate many service people, there is no need for intimidation. The new solid-state control systems are nothing more than a group of circuits used to control a system. Furthermore, there are many similarities between these systems and the old conventional systems.

The heating, air-conditioning, and refrigeration technician must become proficient at troubleshooting the solid-state modules now appearing in the field. The main thrust of the technician's troubleshooting will be aimed at the module rather than each individual solid-state component housed in the module. In most cases, the technician will determine that a solid-state module is faulty and will change the entire module than attempt to troubleshoot the module components. In some cases, electrical devices have been produced that are used to test the more complex electronic modules. Most heating, air-conditioning, and refrigeration companies are not equipped to repair solid-state modules. There, once the determination is made that the module is bad, it is merely replaced with no further testing. With the influx of solid-state modules into the industry, it is imperative that technicians be able to effectively troubleshoot them.

Key Terms

Anti-short-cycling timer
Defrost module
Diode
Electronic module
Electronic timer
Ignition module

Light-emitting diode
Motor protection module
One-function solid-state device
Rectifier
Semiconductor
Solid state

Thermistor
Transistor
Triac
Varistor
Voltage spike

REVIEW TEST

Name: _____   Date: _____   Grade: ___

Answer the following questions

1. What are semiconductors?

2. What is the advantage of using a transistor instead of a vacuum tube?

3. What is a diode?

4. What is a rectifier?

5. What is a transistor?

6. What is the difference between PNP and NPN transistors?

7. What is a one-function solid-state device?

8. What is a multifunction solid-state device?

9. In what applications are solid-state timers used in the industry?

10. How do voltage spikes affect solid-state control circuits?

11. What is the most common use of the LCD in electronic controls?

12. What is a thermistor?

13. What is the purpose of a heat pump electronic module?

14. Why does a heat pump need a means to defrost the outdoor coil?

15. What functions of a gas furnace does an electronic gas furnace control board supervise?

16. What is the advantage of using solid-state motor protection?

17. What part does the cad cell play in the safe operation of an oil-fired furnace?

18. How should a technician approach troubleshooting a single-function solid-state module?

19. How should a technician approach troubleshooting a multifunction electronic module?

20. How is a diagnostic LED on a heat pump or gas furnace?

Name: _____	Date: _____	Grade: ___

Comments:

Objectives: Upon completion of this lab, you should be able to identify single-function solid-state modules and install an electronic timer in a circuit to delay the starting of a motor.

Introduction: Many single-function solid-state modules are used in the air-conditioning, heating, and refrigeration industry. Single-function solid-state controls are used to operate fan motors, defrost heat pumps, and perform other time functions in a controls system. Technicians should be able to install and troubleshoot single-function solid-state controls in HVAC equipment.

Text references: Paragraphs 13.1 through 13.7

Tools and materials: The following materials and equipment will be needed to complete this lab exercise.
    Selection of single-function solid-state modules
    110-Volt power supply
    24-Volt transformer
    Thermostat
    Contactor (24-volt coil)
    Small electric motor
    12″ × 12″ Plywood board

Safety Precaution: Make certain that the electrical source is disconnected when making electrical connections.
• Make sure all connections are tight.
• Make sure no bare current-carrying conductors are touching metal surfaces except the grounding conductor.
• Make sure the correct voltage is being supplied to the circuit.
• Make sure body parts do not come in contact with live electrical conductors.
• Keep hands and materials away from moving parts.

LABORATORY SEQUENCE (mark each box upon completion of task)

A. Identification of Single-function Solid-state Modules

1. Obtain a selection of six single-function solid-state modules from your instructor.

2. On Data Sheet 13A, record the function of each single-function solid-state device.

DATA SHEET 13A

Solid-State module	Function
1.	_____
2.	_____
3.	_____
4.	_____
5.	_____
6.	_____

B. Delay on Make Solid-state Timer

☐   1. Complete (Figure 13.8) so that the small electric motor will start 30 seconds after the thermostat closes, after you have made the proper connections using the pictorial diagram.

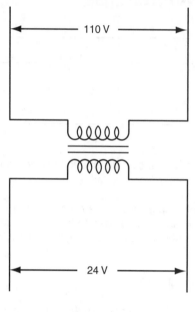

Figure 13.8

☐   2. Have your instructor check the pictorial diagram.

☐   3. Using your diagram, wire the circuit on a plywood board.

☐   4. Connect the circuit to 115-volt power supply.

☐   5. Have your instructor check your circuit.

☐   6. Close the switch for the 115-volt power supply.

☐   7. Close the thermostat.

☐   8. Record what happens when the thermostat closes.

☐   9. Disconnect the power source from the circuit and remove the components from the board.

MAINTENANCE OF WORKSTATION AND TOOLS: Clean and return all tools to their proper locations. Replace all equipment covers. Clean up the workstation.

SUMMARY STATEMENT: What is the action of a solid-state timer when it is energized?

Questions

1. What would be the function in a control circuit of a solid-state timer that delays on closing?

2. What would be the function in a control circuit of a solid-state timer that delays on opening?

3. Why is a defrost control needed for a heat pump in the heating cycle?

4. Name at least five single-function solid-state controls used in the HVAC industry?

5. Draw a schematic diagram of an electrical circuit controlling a fan motor with the starting of the fan motor delayed 60 seconds?

6. Draw a schematic diagram of a heat pump using a single-function solid-state defrost board?

7. What procedure would a technician use to troubleshoot a single-function electronic module?

8. What are the advantages of using solid-state devices control systems?

9. What is a thermistor?

10. What are some reasons that electronic controls modules have made their rapid advancement into the HVAC industry?

LAB 13–2 Multifunction Solid-state Modules

Name: _____	Date: _____	Grade: ___
Comments:		

Objective: Upon completion of this lab, you should be able to identify, understand, identity the inputs and outputs of the module and install a multifunction electronic modules.

Introduction: Many HVAC control systems utilize electronic modules to supervise the operation of heating and air-conditioning equipment. Most heat pumps use electronic modules to supervise their operation. The inputs to a heat pump electronic module would be from the thermostat, defrost thermostat, low-pressure switch, and other controlling elements that the manufacturer deems necessary. These inputs would determine which loads in the system would be required to produce the desired system output. Gas furnaces use electronic modules to supervise the following functions: 1) pilot safety, 2) lockout in the event of flame failure, 3) prepurge, 4) postpurge, 5) blower motor operation, 6) high temperature in furnace, and 7) other functions. Many of these electronic modules are equipped with a diagnostic function that leads the technician to problems when encountered by the equipment.

Text Reference: Paragraphs 13.8 and 13.9

Tools and Materials: The following materials and equipment will be needed to complete this lab exercise.
    Four multifunction electronic modules (two gas furnace modules and two heat pump modules)
    One operating heat pump with electronic module
    One operating gas furnace with electronic module

Safety precautions: Make certain that body parts do not come in contact with live electrical circuits when inspecting modules in operating heat pump and furnace.

LABORATORY SEQUENCE (Mark each box upon completion of task)

A. Input and output identification of electronic modules

1. Obtain four electronic modules from your instructor (two heat pump modules and two gas furnace modules).

2. On Data Sheet 13B, record the type of each module and the inputs and outputs.

DATA SHEET 13B

Module	Type	Inputs	Outputs
#1	_____	_____	_____
		_____	_____
		_____	_____
		_____	_____

#2 _____ _____ _____

   _____ _____

   _____ _____

   _____ _____

#3 _____ _____ _____

   _____ _____

   _____ _____

   _____ _____

#4 _____ _____ _____

   _____ _____

   _____ _____

   _____ _____

B. Operating Electronic Modules (One Heat Pump and One Gas Furnace)

1. Locate and observe the operation of the heat pump with electronic module.

2. Complete Data Sheet 13C

DATA SHEET 13C

Inputs of Board _____

_____

_____

_____

Outputs of Board _____

_____

_____

_____

Function of Board _____

_____

_____

_____

3. Locate and observe the operation of a gas furnace with electronic module.

4. Complete Data Sheet 13D

DATA SHEET 13D

Inputs of Board   _____

_____

_____

_____

Outputs of Board   _____

_____

_____

_____

Function of Board   _____

_____

_____

_____

5. Have your instructor check your data sheets.

MAINTENANCE OF WORK STATION AND TOOLS: Return electronic modules to proper location and clean and return tools to their proper locations. Replace all equipment covers. Clean up work area.

SUMMARY STATEMENT: Why are electronic modules used to supervise the operation of heat pumps and gas furnaces?

Questions

1. What is the purpose of the signal light on an electronic module used on a gas furnace?

2. Why is it important for the technician to know the inputs and outputs to an electronic module used on a heat pump or gas furnace?

3. Why is it impractical for a technician to check the electronic components built into an electronic module?

4. What functions of a gas furnace does the electronic control module supervise?

5. What are the inputs and outputs of the heat pump electronic control module shown in (Figure 13.9)?

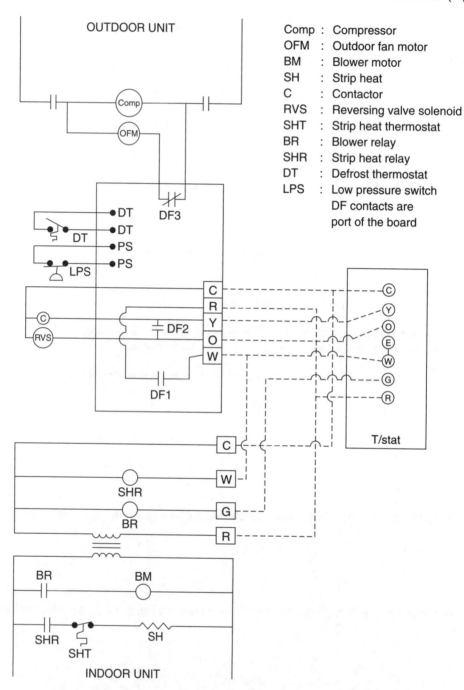

Comp	:	Compressor
OFM	:	Outdoor fan motor
BM	:	Blower motor
SH	:	Strip heat
C	:	Contactor
RVS	:	Reversing valve solenoid
SHT	:	Strip heat thermostat
BR	:	Blower relay
SHR	:	Strip heat relay
DT	:	Defrost thermostat
LPS	:	Low pressure switch
		DF contacts are
		port of the board

Figure 13.9 Diagram of a heat pump with control module.

6. What are the inputs and outputs of the gas furnace electronic control module shown in (Figure 13.10)?

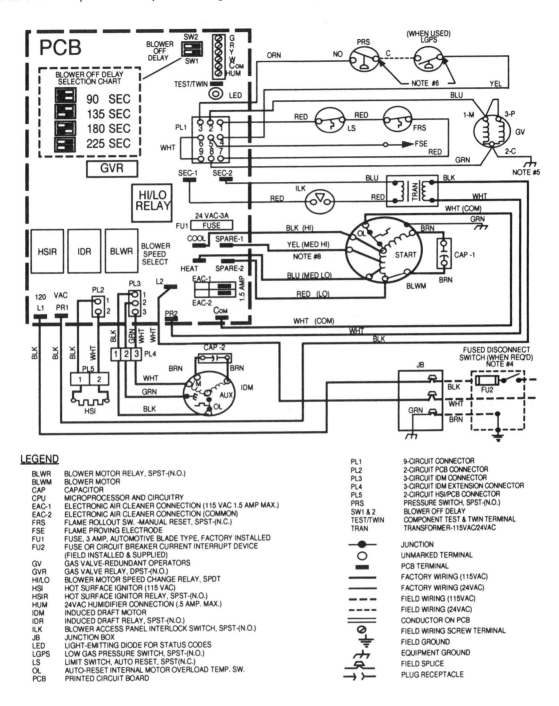

Figure 13.10 Diagram of gas furnace with electronic control module.

LEGEND

BLWR	BLOWER MOTOR RELAY, SPST-(N.O.)
BLWM	BLOWER MOTOR
CAP	CAPACITOR
CPU	MICROPROCESSOR AND CIRCUITRY
EAC-1	ELECTRONIC AIR CLEANER CONNECTION (115 VAC 1.5 AMP MAX.)
EAC-2	ELECTRONIC AIR CLEANER CONNECTION (COMMON)
FRS	FLAME ROLLOUT SW. -MANUAL RESET, SPST-(N.C.)
FSE	FLAME PROVING ELECTRODE
FU1	FUSE, 3 AMP, AUTOMOTIVE BLADE TYPE, FACTORY INSTALLED
FU2	FUSE OR CIRCUIT BREAKER CURRENT INTERRUPT DEVICE (FIELD INSTALLED & SUPPLIED)
GV	GAS VALVE-REDUNDANT OPERATORS
GVR	GAS VALVE RELAY, DPST-(N.O.)
HI/LO	BLOWER MOTOR SPEED CHANGE RELAY, SPDT
HSI	HOT SURFACE IGNITOR (115 VAC)
HSIR	HOT SURFACE IGNITOR RELAY, SPST-(N.O.)
HUM	24VAC HUMIDIFIER CONNECTION (.5 AMP. MAX.)
IDM	INDUCED DRAFT MOTOR
IDR	INDUCED DRAFT RELAY, SPST-(N.O.)
ILK	BLOWER ACCESS PANEL INTERLOCK SWITCH, SPST-(N.O.)
JB	JUNCTION BOX
LED	LIGHT-EMITTING DIODE FOR STATUS CODES
LGPS	LOW GAS PRESSURE SWITCH, SPST-(N.O.)
LS	LIMIT SWITCH, AUTO RESET, SPST(N.C.)
OL	AUTO-RESET INTERNAL MOTOR OVERLOAD TEMP. SW.
PCB	PRINTED CIRCUIT BOARD

PL1	9-CIRCUIT CONNECTOR
PL2	2-CIRCUIT PCB CONNECTOR
PL3	3-CIRCUIT IDM CONNECTOR
PL4	3-CIRCUIT IDM EXTENSION CONNECTOR
PL5	2-CIRCUIT HSI/PCB CONNECTOR
PRS	PRESSURE SWITCH, SPST-(N.O.)
SW1 & 2	BLOWER OFF DELAY
TEST/TWIN	COMPONENT TEST & TWIN TERMINAL
TRAN	TRANSFORMER-115VAC/24VAC

●	JUNCTION
○	UNMARKED TERMINAL
▬	PCB TERMINAL
───	FACTORY WIRING (115VAC)
───	FACTORY WIRING (24VAC)
---	FIELD WIRING (115VAC)
---	FIELD WIRING (24VAC)
═══	CONDUCTOR ON PCB
⊘	FIELD WIRING SCREW TERMINAL
⏚	FIELD GROUND
⏚	EQUIPMENT GROUND
⟿	FIELD SPLICE
⟶	PLUG RECEPTACLE

7. How does a cad cell prove that an oil furnace has ignited?

8. If a technician determines that an electronic module is faulty on a heat pump or gas furnace, what action should they take?

9. What troubleshooting procedure would a technician use to determine if a multifunction electronic module was functioning properly?

10. How has the advancement of electronic modules changed control systems on heat pumps and gas furnaces?

CHAPTER 14 Heating Control Devices

Chapter Overview

In the winter months, most residences and other occupied structures will require an adequate supply of heat to maintain a comfortable temperature. The three major sources of energy used to supply heat to a structure are gas (natural or liquefied petroleum), oil, and electricity. These sources of energy can be used to heat air in a warm air heating system, heat water in a hydronic heating system, or produce steam for a steam heating system. Many of the controls used in heating systems are similar to controls that have been previously discussed, such as thermostats, pressure switches, and relays. A thermostat can be used as a safety switch to interrupt electrical power going to a device that controls the heating source if the appliance overheats (Figure 14.1) or start a fan motor in a furnace when the combustion chamber is warm enough to supply heat to the structure (Figure 14.2). The function of the controls can be determined by examination of the schematic diagram.

One of the primary purposes of heating control devices is to supervise the supply of the heating source to the furnace or boiler. Gas heating equipment requires one of the following methods to insure that the pilot is available or that there is a method of ignition for the main burner: a standing pilot, intermittent ignition, or direct ignition. The intermittent ignition pilot system and the direct spark ignition system have gained popularity over the past ten years because of the added efficiency and the added safety features along with the decrease in cost.

The old standing pilot system uses a gas valve with a pilot solenoid that will hold the pilot valve open once a pilot flame has been established. The pilot solenoid is held in once depressed by the millivoltage produced by a thermocouple, which is accomplished by the pilot flame contacting the thermocouple. A thermocouple and pilot assembly is shown in (Figure 14.3). If the millivoltage produced by the thermocouple drops to an unacceptable level, the pilot valve will close and stop any gas from passing through the main gas valve (Figure 14.4). The intermittent ignition system and the direct spark ignition system both use some type of electronic ignition mod-

Figure 14.1 Flame rollout limit switch. *(Courtesy of Therm-O-Disc, Mansfield, OH)*

Figure 14.2 Combination fan and limit switch. *(Courtesy of Honeywell, Inc.)*

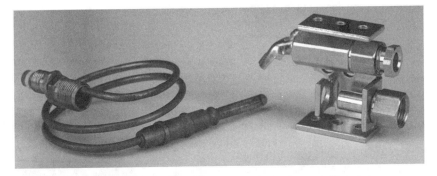

Figure 14.3 Photograph of a thermocouple and pilot burner. *(Courtesy of Honeywell, Inc.)*

Figure 14.4 Photograph of a gas valve. *(Courtesy of Honeywell, Inc.)*

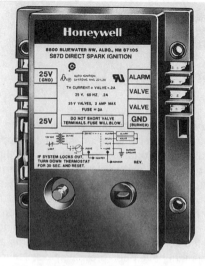

Figure 14.6 Photograph of ignition module used on a direct ignition control system. *(Courtesy of Honeywell, Inc.)*

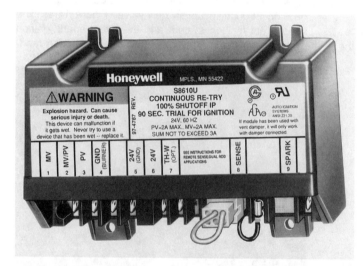

Figure 14.5 Photograph of ignition module. *(Courtesy of Honeywell, Inc.)*

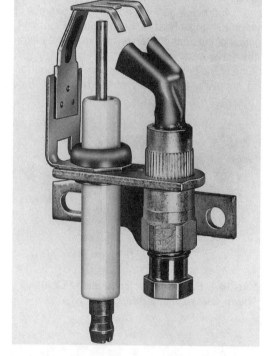

Figure 14.7 Photograph of a pilot assembly and flame rod. *(Courtesy of Honeywell, Inc.)*

ule to supervise the ignition of the pilot or main burner. An intermittent pilot ignition module is shown in (Figure 14.5) and a direct spark ignition module is shown in (Figure 14.6). The pilot in these two types of systems is ignited by a spark or a hot surface igniter. An ignition module used with spark ignition is shown in (Figure 14.7). A hot surface igniter is shown in (Figure 14.8).

There are two types of oil burners in use in the industry today, vaporizing and atomizing. In general, before fuel oil can be ignited it must be vaporized or broken down into a fine mist and mixed with air. The vaporizing oil burner depends on natural evaporation facilitated by the heating of the fuel oil to provide oil vapor for combustion. A good example of this method is the carburetor used on older-model furnaces. The atomizing oil burner uses a pump and fan connected to the burner motor to accomplish this

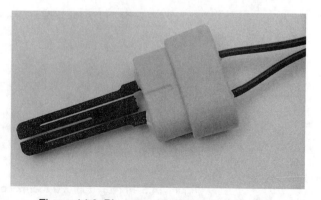

Figure 14.8 Photograph of hot surface igniter.

Figure 14.9 Photograph of stack switch. *(Courtesy of Honeywell, Inc.)*

Figure 14.11 Cad cell primary control. *(Courtesy of Honeywell, Inc.)*

purpose. The oil pump supplies oil from the storage tank and raises the pressure of the oil entering the oil burner nozzle, causing the oil to be supplied to the combustion chamber in a fine mist. The oil burner fan causes the air to mix with the oil and is ignited by a spark from an ignition transformer. There are two types of combustion controls commonly used on residential oil-fired heating equipment, the stack switch and the cad cell. The stack switch is shown in (Figure 14.9) and senses the heat in the stack of oil-fired heating equipment. The cad cell shown in (Figure 14.10) senses the light from the ignited oil and sends a signal to the cad cell primary control shown in (Figure 14.11) to insure that ignition has been proven.

Heating systems that use electrical energy as the heating source use many of the same controls that have already been covered in this lab manual. The sequencer is a control that is used in electric heating equipment. A sequencer is shown in (Figure 14.12) and its diagram is shown in (Figure 14.13). The sequencer is nothing more than a time-delay relay that closes the contacts at intervals to prevent a high current draw at one time.

There are many thermostats, pressure switches, and other common electrical components that are used in heating systems. The technician can usually determine the function of these electrical devices by referring to the schematic diagram for the equipment.

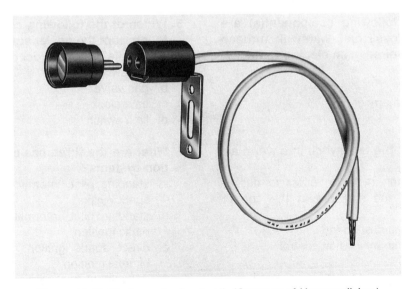

Figure 14.10 Photograph of cad cell. *(Courtesy of Honeywell, Inc.)*

Figure 14.12 Photograph of sequencer. *(Courtesy of Honeywell, Inc.)*

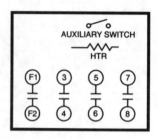

Figure 14.13 Diagram of sequencer. *(Courtesy of Honeywell, Inc.)*

Key Terms

Electrical resistance heater
Fan switch
Gas valve
Hot surface igniter

Hot surface ignition
Ignition module
Pilot
Pilot assembly

Primary control
Sequencer
Spark igniter
Thermocouple

REVIEW TEST

Name: _____   Date: _____   Grade: ___

Complete the following multiple choice questions by selecting the correct answer.

1. **What are the common energy sources used for heating?**
   a. gas
   b. oil
   c. electricity
   d. all of the above

2. **Which of the following component(s) are required in a fossil fuel, warm air furnace and are not required in an electric furnace?**
   a. fan
   b. limit switches
   c. combustion chamber
   d. fan control

3. **The purpose of the fan switch in a warm air furnace is to _____.**
   a. control the fan motor in order to deliver warm air to the structure at the correct temperature
   b. control the combustion fan
   c. control the primary ignition control
   d. all of the above

4. **A time-controlled fan switch is nothing more than a _____.**
   a. time clock
   b. time-delay relay
   c. bimetal element
   d. mechanical timer

5. **Which of the following components is used to interrupt the power source to a load when an unsafe condition occurs?**
   a. fan switch
   b. gas valve
   c. time clock
   d. limit switch

6. **What are the three basic types of gas ignition systems?**
   a. standing pilot, intermittent pilot, and direct spark ignition
   b. standing pilot, intermittent pilot, and automatic ignition
   c. direct spark ignition, standing pilot, and bimetal ignition
   d. intermittent pilot, direct spark ignition, and subsurface ignition

7. What component in a standing pilot ignition system produces the signal that a pilot is available?
   a. thermocouple
   b. flame rod
   c. bimetal element
   d. hot surface igniter

8. Which of the following types of flame sensors could not be used with an intermittent pilot ignition system?
   a. flame rod
   b. liquid-filled pilot sensor
   c. temperature sensor
   d. thermocouple

9. What type of gas valve is used with an intermittent pilot ignition system?
   a. combination
   b. direct
   c. redundant
   d. combustion

10. A direct ignition system _____.
    a. lights the pilot
    b. lights the main burner
    c. both a and c
    d. none of the above

11. What ignites the atomized fuel oil in a residential oil burner?
    a. pilot
    b. high-voltage spark
    c. hot surface igniter
    d. none of the above

12. What would be the results if the oil burner were allowed to run without ignition?
    a. explosion
    b. delayed ignition
    c. accumulation of fuel oil in the combustion chamber
    d. none of the above

13. Which of the following is a light-sensitive device that changes its resistance according to the intensity of the light?
    a. thermocouple
    b. fuel limiter cell
    c. subradiant cell
    d. cad cell

14. What type of primary control used on an oil burner uses temperature to prove ignition?
    a. stack switch
    b. fan switch
    c. limit switch
    d. photo cell

15. What electrical control device is used in an electric furnace to bring in electric resistance heaters in stages?
    a. contactor
    b. line voltage thermostats
    c. sequencers
    d. none of the above

16. Which of the following types of heating appliances are used to heat water or produce steam?
    a. boilers
    b. furnaces
    c. elements
    d. none of the above

17. Which of the following components would be used in the control system of a steam boiler?
    a. pressure switch
    b. limit switches
    c. low water cut-off
    d. all of the above

18. The resistance of the cad cell should be below _____ ohms to continue operation of an oil burner.
    a. 35,000
    b. 2500
    c. 1600
    d. 300

19. What is the proper location of a stack switch for proper supervision of the oil burner?
    a. combustion chamber
    b. oil burner stack
    c. oil burner flame
    d. blower compartment

20. A limit switch used on an oil- or gas-fired warm air furnace would usually _____ on a temperature rise.
    a. open
    b. close

LAB 14-1 Electric Heating Controls

Name: _____   Date: _____   Grade: ___

Comments:

Objectives: Upon completion of this lab, you should be able to draw, make the electrical connections on a practice wiring board, and wire an electric furnace.

Introduction: Electric heating has lost popularity over the last several years due to its high energy cost. Electric resistance heat is still being used occasionally as the primary heating source, but more often it is used as the supplementary or emergency heat source on heat pump applications. No matter where the technician finds these applications, there are many similarities in the control systems. When electricity is used as the primary heat source, sequencers are used to control the resistance heaters by staging the starting and stopping of the heaters. In heat pump applications, the control of electric resistance heaters is by the indoor and outdoor thermostats, if the system is so equipped. The heaters and safety controls will all be very similar in design.

Text Reference: Paragraph 14.5

Tools and Materials: The following materials and equipment will be needed to complete this lab exercise.
Electric furnace with installation instructions
Sequencers
Transformer
Relays
Thermostat
Other miscellaneous electrical devices
Volt-ohmmeter
Clamp-on ammeter
Plywood to be used for practice wiring board
Miscellaneous wiring supplies
Basic electrical handtools

Safety Precautions: Make certain that the electrical source is disconnected when making electrical connections. In addition:
- Make sure all connections are tight.
- Make sure no bare current-carrying conductors are touching metal surfaces except the grounding conductor.
- Make sure the correct voltage is being supplied to the circuits.
- Make sure body parts do not come in contact with live electrical conductors.
- Keep hands and materials away from moving parts.

LABORATORY SEQUENCE (mark each box upon completion of task)

A. Electric Resistance Heating Circuitry

☐ 1. On the following page, draw a schematic diagram for an electric furnace to be operated on 115 volts with two electric resistance heaters and a blower motor. Include the following safety components:
- Fuses to protect the heaters from current overload
- Temperature limits for each heater to insure that overheating does not occur

☐ 2. Have your instructor check your wiring diagram.

☐ 3. Compile a list of the components and materials that will be needed to install the circuits on the practice board. (NOTE: Use light bulbs to represent the electric heaters and blower motor.)

☐ 4. Obtain the necessary components and supplies from your instructor or supply room.

☐ 5. Install the components on the practice board and make the necessary electrical connections to complete your diagram.

☐ 6. Have your instructor check your wiring board.

☐ 7. Connect the wiring board to a 115-volt power source.

☐ 8. Operate the control system.

☐ 9. Record the operational sequence of the electric heating system.

☐ 10. Disconnect the electrical board from the power source.

☐ 11. Remove the electrical components from the board and return them to their proper location.

B. Installation of an Electric Furnace

☐ 1. Obtain a unit assignment (electric furnace) from your instructor.

☐ 2. Read the installation instructions for the electric furnace.

☐ 3. Determine and record the correct wire size for the electric furnace.

   Wire size _____

☐ 4. Compile a list of materials that will be needed to install the assigned electric furnace.

☐ 5. Make the necessary electrical connections for the proper installation of the electric furnace.

☐ 6. Have your instructor check your installation.

☐ 7. Operate the electric furnace and check the following electrical characteristics of the system.

   Supply voltage _____

   Control voltage _____

   Operating current _____

☐ 8. Write a brief explanation of the operation of the electric furnace.

☐ 9. Follow the instructor's directions for disassembling your installation.

MAINTENANCE OF WORK STATION AND TOOLS: Clean and return all tools to their proper location(s). Replace all equipment covers. Clean up the work area.

SUMMARY STATEMENT: Why are sequencers commonly used instead of contactors to control the electric resistance heaters in an electric furnace?

Questions

1. What has caused the decrease in popularity of the electric furnace?

2. How are supplementary electric resistance heaters controlled in a heat pump system?

3. How can fuses adequately protect electric resistance heaters?

4. Draw the diagram of a sequencer that would be used in an electric furnace to control three electric resistance heaters and the fan motor.

5. If fuses protect the heaters, why are limit switches needed in electric resistance heater applications?

6. What is the Btu/hr output of a 5-kW resistance heater?

7. Draw a diagram of an electric furnace with three electric resistance heaters and a fan motor; include all necessary safety components.

8. What is the advantage of using sequencers in electric furnaces?

9. What is the difference between an electric furnace and a supplementary heater used in heat pump installations?

10. Why is it important to have some type of fan interlock on electric resistance heater applications?

Name: _____	Date: _____	Grade: ___

Comments:

Objectives: Upon completion of this lab, you should be able to correctly install and wire gas controls on gas-fired, warm air furnaces in the shop.

Introduction: There are several types of ignition controls used on gas-fired, warm air furnaces: standing pilot, intermittent pilot, and direct spark ignition. The technician will find all three types of ignition control systems in the field and should understand the operation of each. The electrical circuitry of most gas furnaces is similar with the exception of the ignition control system. Many gas furnaces are now equipped with some type of vent fan, which also must be controlled by the control system.

Text Reference: Paragraphs 14.2 and 14.3

Tools and Materials: The following materials and equipment will be needed to complete this lab exercise.

Gas furnace with standing pilot ignition system   Hot surface igniter
Gas furnace with intermittent ignition system   Intermittent ignition module
Gas furnace with direct ignition system   Direct ignition module
Gas valves   Volt-ohmmeter
Thermocouples   Basic electrical handtools
Pilot burners   Tools to remove gas valve
Pilot assemblies   Miscellaneous wiring supplies

Safety Precautions: Make certain that the electrical source is disconnected when making electrical connections. In addition:

- Make sure all connections are tight.
- Make sure no bare current-carrying conductors are touching metal surfaces except grounding conductors.
- Make sure the correct voltage is being supplied to the circuits or equipment.
- Make sure body parts do not come in contact with live electrical conductors.
- Keep hands and materials away from moving parts.

When installing components in gas lines, make sure there are no leaks. When igniting gas furnaces, do not stand directly in front of the combustion chamber. Do not move ignition control components from the manufacturer's location. Make sure the ignition control is correct for the application in which it is being used.

LABORATORY SEQUENCE (mark each box upon completion of task)

A. Gas Furnace Circuitry

☐   1. Obtain a furnace assignment from your instructor.

☐   2. Make sure the power is off and the electrical switch is adequately marked "SERVICE IN PROGRESS."

☐   3. Remove the front cover(s) of the furnace and the junction box cover(s).

☐   4. Examine the components and wiring of the furnace. If you do not understand any component and its operation, refer to your text; if further assistance is needed, consult with your instructor. Once you completely understand the wiring and operation of the furnace, draw a pictorial diagram and a schematic diagram in the space provided on the following page. Complete the line voltage and low-voltage sections of the furnace. Do not include a thermocouple in either drawing.

☐   5. Replace the blower compartment cover.

Pictorial diagram

Schematic diagram

☐ 6. Operate the furnace and briefly explain in writing the function of each component and the control sequence.

☐ 7. Have your instructor check your diagrams and explanation of furnace operation.

☐ 8. Replace all covers and panels.

B. Standing Pilot Ignition System, Millivolt Production of Thermocouple

☐ 1. Obtain a furnace assignment from your instructor.

☐ 2. Turn off the power supplying the furnace and tag the switch with a "SERVICE IN PROGRESS" tag.

☐ 3. Remove front panel of furnace and panel covering the combustion chamber.

☐ 4. Install a thermocouple adapter where the thermocouple is attached to the gas valve.

☐ 5. Light the furnace pilot. (NOTE: Follow the manufacturer's instructions for lighting the pilot.)

☐ 6. Record the millivoltage produced by the thermocouple when the pilot dial is depressed (unloaded) and when the pilot dial is released (loaded).

Millivoltage (unloaded) _____

Millivoltage (loaded) _____

☐ 7. Restore power to the furnace and set the thermostat to a call for heat.

☐ 8. Record the millivoltage produced by the thermocouple with the main burner operating.

_____

☐ 9. Replace all covers and panels.

C. Changing a Gas Valve on a Furnace with a Standing Pilot

☐ 1. Obtain a gas furnace assignment from your instructor along with a replacement gas valve.

☐ 2. Turn off the power supplying the furnace and tag the switch with a "SERVICE IN PROGRESS" tag. Close the gas supply to the furnace.

☐ 3. Remove the panels from the furnace covering the burner compartment and combustion chamber.

☐ 4. Draw a wiring diagram of the wiring that must be removed. Remove the wires from the gas valve. Take care in removing wiring from the gas valve.

☐ 5. Examine the gas piping and determine the disassembly procedure. Look for a union or flare connection at the inlet of the gas valve.

☐ 6. Remove the pilot line connections from the gas valve. Use the correct tools for the job.

☐ 7. Disassemble the piping to the gas valve. Use the right tools for the job.

☐ 8. Examine the gas valve application to determine the type of pilot ignition. Pay careful attention to where the ignition components connect. Remove the ignition components from the gas valve.

☐ 9. Remove the gas valve from the gas manifold.

☐ 10. Install the gas valve obtained from your instructor back to the gas manifold. Use joint compound or Teflon tape on all external pipe threads.

☐ 11. Connect the gas supply back to the inlet of the gas valve.

☐ 12. Connect the ignition components back to the gas valve.

☐ 13. Make all electrical connections on the gas valve. Make sure connections are correct; refer to your diagram.

☐ 14. Have your instructor check your gas valve installation.

☐ 15. Turn the gas supply on.

☐ 16. Check the gas supply piping for leaks using soap bubbles.

☐ 17. Restore the power supply to the furnace.

☐ 18. Light the pilot if the furnace has a standing pilot ignition system.

☐ 19. Set the thermostat to a call for heat. When ignition of the burner is accomplished, immediately check the remaining piping for leaks, including the pilot lines.

☐ 20. When you have determined that no leaks exist, turn off the furnace.

☐ 21. Clean soap solution and residue from gas valve and piping.

☐ 22. Replace all panels on the furnace.

D. Intermittent Pilot Ignition System

☐ 1. Obtain a furnace assignment from your instructor.

☐ 2. Assume that the intermittent ignition module on the assigned furnace is faulty and must be replaced.

☐ 3. Remove any furnace panels necessary to obtain access to the ignition module.

☐ 4. Perform preinstallation safety inspection.
   a. Test gas piping for gas leaks.
   b. Visually inspect venting system for proper installation and size.
   c. Inspect burners and crossovers for blockage and corrosion.
   d. Inspect heat exchanger in warm air furnace for damage such as cracks and excessive corrosion. If a boiler, check for water leaks and combustion gas leaks.
   e. Make any other safety checks required by state and local codes.

☐ 5. Obtain a new intermittent ignition module from your instructor.

☐ 6. Compare the rating of the old ignition module with the new ignition module to make sure it is suitable for your application.

☐ 7. Disconnect the power supply and tag the switch "SERVICE IN PROGRESS" before doing any work on the unit.

☐ 8. Disconnect and tag the wires from the old module.

☐ 9. Remove the old module from its mounting location.

☐ 10. Mount the new module in the same location as the old module, if possible.

☐ 11. Wire the module, checking the wiring diagram furnished by the manufacturer and the module installation instructions.

☐ 12. Verify the thermostat heat anticipator setting.

☐   13. Visually inspect the module installation, making sure all wiring connections are clean and tight.

☐   14. Verify the control system ground. The igniter, flame sensor, and ignition module must share a common ground with the main burner.

☐   15. Have your instructor check your wiring.

☐   16. Restore power to the furnace.

☐   17. Review the normal operating sequence and module specifications.

☐   18. Check the safety shutoff operation. Turn off the gas supply. Set the thermostat to a call for heat. Watch for a spark at the pilot burner and time the spark from start to shutoff; this time should be 90 seconds maximum. Open the manual gas valve and make sure no gas is flowing to the pilot or main burner.

☐   19. Set the thermostat to the lowest setting and wait one minute.

☐   20. Check the normal operation of the ignition module by moving the thermostat to a call for heat. Make sure the pilot lights smoothly. Make sure the main burner lights smoothly without flashback and that the main burner is correctly adjusted.

☐   21. Return the thermostat to a setting below room temperature. Make sure the main burner goes out.

☐   22. Replace all furnace panels.

☐   23. Return replaced ignition module to your instructor.

☐   24. Clean up the work area.

E. Direct Ignition System

☐   1. Obtain a furnace assignment from your instructor.

☐   2. Assume that the direct ignition module on the assigned furnace is faulty and must be replaced.

☐   3. Remove any furnace panels necessary to obtain access to the ignition module.

☐   4. Perform preinstallation safety inspection.
      a. Test gas piping for gas leaks.
      b. Visually inspect venting system for proper installation and size.
      c. Inspect burners and crossovers for blockage and corrosion.
      d. Inspect heat exchanger in warm air furnace for damage such as cracks and corrosion. If a boiler, check for water leaks and combustion gas leaks.
      e. Make any other safety checks required by state and local codes.

☐   5. Obtain a new direct ignition module from your instructor.

☐   6. Compare the rating of the old ignition module with the new ignition module and make sure it is suitable for your application.

☐   7. Disconnect the power supply and tag the switch "SERVICE IN PROGRESS" before doing any work on the unit.

☐   8. Disconnect and tag the wires from the old module.

☐   9. Remove the old module from its mounting location.

☐ 10. Mount the new module in the same location as the old module, if possible.

☐ 11. Wire the module, checking the wiring diagram furnished by the manufacturer and the module installation instructions.

☐ 12. Verify the thermostat heat anticipator setting.

☐ 13. Visually inspect the module installation, making sure all wiring connections are clean and tight.

☐ 14. Verify the control system ground. The igniter, flame sensor, and ignition module must share a common ground with the main burner.

☐ 15. Have your instructor check your wiring.

☐ 16. Restore power to the furnace.

☐ 17. Review the normal operating sequence and module specifications.

☐ 18. Set the thermostat to a temperature setting higher than room temperature for at least one minute.

☐ 19. Check safety shutoff operation. First, turn the gas supply off and set the thermostat above the room temperature to a call for heat. Observe the operation of the spark igniter or hot surface igniter. (A spark igniter will spark following prepurge and a hot surface igniter will begin to glow several seconds after prepurge.)

Determine the trial ignitions from the ignition module installation instructions to determine the number of trial ignitions that the control should allow. Open the manual knob on the gas valve and make sure no gas flows to the burner. Set the thermostat below room temperature.

☐ 20. Check the normal operation of the ignition module by setting the thermostat above room temperature. Observe the lighting sequence and make sure the main burner lights smoothly and without flashback. Make sure the burner operates smoothly without floating, lifting, or flame rollout.

☐ 21. Return the thermostat to a setting below room temperature. Make sure the main burner goes out.

☐ 22. Replace all furnace panels.

☐ 23. Return replaced ignition module to your instructor.

☐ 24. Clean up the work area.

MAINTENANCE OF WORK STATION AND TOOLS: Clean and return all tools to their proper location(s). Replace all equipment covers. Clean up the work area.

SUMMARY STATEMENT: What is the difference between a standing pilot ignition system, an intermittent ignition system, and a direct ignition system?

Questions

1. What is the purpose of a limit switch in the control system of a gas furnace?

2. What is a thermocouple?

3. Explain the operation of an intermittent ignition module.

4. What should be the millivolt output of a good thermocouple?

5. What is a hot surface igniter?

6. What is the purpose of a prepurge cycle in a direct ignition system?

7. Why should the safety lockout of gas ignition systems always be checked when new components are installed?

8. What is the response time between the lockout of a standing pilot ignition system and an intermittent pilot ignition system?

9. Why should a technician always label the wiring when removing wires from an ignition module that is to be replaced?

10. Why is it important for a technician always to replace ignition components in the location where they were placed by the manufacturer?

LAB 14–3 Oil Heating Controls

Name: _____   Date: _____   Grade: ____

Comments:

Objectives: Upon completion of this lab, you should be able to correctly install oil burner primary controls on oil-fired, warm air furnaces in the shop.

Introduction: Oil-fired, warm air furnaces are equipped with primary controls that turn the burner off and on in response to the thermostat action and monitor the oil burner flame. There are basically two types of primary controls used in the industry, the stack switch and the cad cell control. The stack switch uses the heat of the flue gases to monitor and prove that a flame has been established, while the cad cell primary control uses a light-sensitive cad cell to prove and monitor the flame of an oil burner. The primary control is the heart of the oil burner control system, but other controls are also important to the safe operation of an oil burner, such as fan and limit switches.

Text Reference: Paragraph 14.4

Tools and Materials: The following materials and equipment will be needed to complete this lab exercise.
Oil-fired furnace
Stack switch primary oil burner controls
Cad cell primary oil burner controls
Volt-ohmmeter
Basic electrical handtools
Miscellaneous wiring supplies

Safety Precautions: Make sure that the electrical source is disconnected when making electrical connections. In addition:
- Make sure all connections are tight.
- Make sure no bare current-carrying conductors are touching metal surfaces except the grounding conductor.
- Make sure the correct voltage is being supplied to the circuits or equipment.
- Make sure body parts do not come in contact with live electrical conductors.
- Keep hands and materials away from moving parts.

When igniting oil furnaces, do not stand directly in front of the combustion chamber access door. Make certain that there are no leaks in the oil piping. Make sure the oil burner primary control is the correct type for the application in which it is being used.

LABORATORY SEQUENCE (mark each box upon completion of task)

A. Stack Switch Oil Burner Primary Controls

☐ 1. Obtain a furnace assignment from your instructor.

☐ 2. Assume that the stack switch primary control on the assigned oil-fired furnace is faulty and must be replaced.

☐ 3. Remove any furnace panels necessary to obtain access to the oil burner and controls.

☐ 4. Perform preinstallation safety inspection.
   a. Make sure oil piping has no leaks.
   b. Visually inspect venting system for proper installation and size.
   c. Inspect oil burner for air blockage and corrosion.
   d. Inspect heat exchanger in warm air furnace for damage such as cracks and corrosion. If a boiler, check for water leaks and combustion gas leaks.
   e. Make any other safety checks required by state and local codes.

☐ 5. Obtain a new stack switch primary control from your instructor.

☐ 6. Compare the rating of the old stack switch with the new stack switch and make sure it is suitable for your application.

☐ 7. Disconnect the power supply and tag the switch "SERVICE IN PROGRESS" before doing any work on the unit.

☐ 8. Disconnect and tag the wires from the stack switch.

☐ 9. Remove the old stack switch from its mounting location.

☐ 10. Mount the new stack switch in the same location as the old stack switch.

☐ 11. Wire the stack switch, checking the wiring diagram furnished by the appliance manufacturer and the stack switch installation instructions.

☐ 12. Verify the thermostat heat anticipator setting.

☐ 13. Visually inspect the stack switch installation, making sure all wiring connections are clean and tight.

☐ 14. Have your instructor check your wiring.

☐ 15. Restore power to the furnace.

☐ 16. Review the normal operating sequence and stack switch specifications. If the new stack switch will not close, put the stack switch in step by pulling the drive shaft lever forward 1/4 inch, then slowly releasing.

☐ 17. Set the thermostat to a temperature setting lower than room temperature for at least one minute.

☐ 18. Close the thermostat and check the safety shutoffs of the stack switch.
    a. Check the flame failure function by shutting off the oil supply to the burner. The stack switch should lock the oil burner out on safety. This safety switch must be reset manually.
    b. Check the scavenger timing of the stack switch by operating the burner normally, then open and immediately close the line voltage switch. The oil burner should stop immediately, and after recycling time (usually 1 to 3 minutes) it should restart automatically.

☐ 19. Check for normal operation of the stack switch by setting the thermostat above room temperature. Observe the lighting sequence and make sure the burner lights smoothly. Make sure the burner operates smoothly with the proper flame.

☐ 20. Return the thermostat setting to below room temperature. Make sure the oil burner goes out.

☐ 21. Replace all furnace panels.

☐ 22. Return the faulty stack switch to your instructor.

☐ 23. Clean up the work area.

B. Cad Cell Oil Burner Primary Controls

☐ 1. Obtain a furnace assignment from your instructor.

☐ 2. Assume that the cad cell primary control on the assigned furnace is faulty and must be replaced.

☐ 3. Remove any furnace panels necessary to obtain access to the oil burner and controls.

4. Perform preinstallation safety inspection.
   a. Make sure oil piping has no leaks.
   b. Visually inspect venting system for proper installation and size.
   c. Inspect oil burner for air blockage and corrosion.
   d. Inspect heat exchanger in warm air furnace for damage such as cracks and corrosion. If a boiler, check for water leaks and combustion gas leaks.
   e. Make any other safety checks required by state and local codes.

5. Obtain a new cad cell and cad cell primary control from your instructor.

6. Plug the cad cell into the holder. Aim the cad cell toward a bright light; measure and record the resistance of the cad cell. Next, cover the cad cell; measure and record the resistance.

   Lighted cad cell resistance _____ ohms

   Dark cad cell resistance _____ ohms

7. Compare the rating of the old cad cell primary control with the new cad cell primary control and make sure it is suitable for your application.

8. Disconnect the power supply and tag the switch "SERVICE IN PROGRESS" before doing any work on the unit.

9. Disconnect and tag the wires from the cad cell primary control.

10. Remove the cad cell primary control and cad cell from its mounted location.

11. Mount the new cad cell and cad cell primary control in the same location as the old cad cell and cad cell primary control. (NOTE: Do not move location of cad cell.)

12. Wire the cad cell primary control, checking the wiring diagram furnished by the unit manufacturer and the cad cell primary control installation instructions.

13. Verify the thermostat heat anticipator setting.

14. Visually inspect the cad cell and cad cell primary control installation and make sure all wiring connections are clean and tight.

15. Have your instructor check your wiring.

16. Restore power to the furnace.

17. Review the normal operating sequence and cad cell primary control specifications.

18. Set the thermostat to a temperature setting lower than room temperature for at least one minute.

19. Close the thermostat and check the safety shutoff of the cad cell primary control by stopping the oil flow to the oil burner. The cad cell primary control should lock out in switch timing (15 to 70 seconds). The oil burner should stop. Most cad cell primary controls must be manually reset.

20. Check for normal operation of the stack switch by setting the thermostat above room temperature. Observe the lighting sequence and make sure the oil burner lights. Make sure the burner operates smoothly with the proper flame.

21. Return the thermostat setting to below room temperature. Make sure the oil burner goes out.

22. Replace all furnace panels.

☐ 23. Return the faulty cad cell and cad cell primary control to your instructor.

☐ 24. Clean up the work area.

MAINTENANCE OF WORK STATION AND TOOLS: Clean and return all tools to their proper location(s). Replace all equipment covers. Clean up the work area.

SUMMARY STATEMENT: What is the purpose of an oil burner primary control?

Questions

1. What senses combustion when a cad cell primary control is used?

2. What senses combustion when a stack switch is used as the primary control?

3. Where is the cad cell located in an oil burner when used with a cad cell primary control?

4. What would be the results if the cad cell was not placed in line with the oil burner flame?

5. Explain the operation of a cad cell primary control.

6. Explain the operation of a stack switch primary control.

7. Where should the element of a stack switch be installed?

8. Why should a service technician check the safety functions of a primary control?

9. What would a service technician do if a new stack switch is installed and will not work?

10. When an oil burner does not ignite the oil being supplied to the combustion chamber, why is it important for the primary control to stop the oil burner?

CHAPTER 15 Troubleshooting Electric Control Devices

Chapter Overview

One of the most important tasks performed by a heating, refrigeration, and air-conditioning technician is the diagnosis of electrical components in the electrical circuitry of equipment and control systems. Therefore, it is imperative for service technicians to understand how electrical components operate, to know how to use electric meters to check components, and to know the proper procedures for checking electric components. Most problems in an HVAC system stem from one source—an electric component that is not functioning properly or is faulty. It is the responsibility of the service technician to locate the component that is not functioning properly or is faulty and to repair or replace that component. Sometimes it is difficult to locate the exact trouble in the system, but the task should be relatively simple once the problem has been narrowed down to a single component or several components.

The first step in troubleshooting any electrical component is to understand its operation and function. If the operation of an electrical component is not understood, it is impossible to effectively check the component. Service technicians must understand the operation of the electrical components and the proper procedure for checking electrical components to correctly diagnose the condition of the component. The service technician must be able to use all types of electrical meters to determine the condition of electrical components. Not only must technicians know how to use electrical meters, they will also have to know the approximate value of the electrical characteristic being tested, such as the expected resistance reading of a relay or contactor coil and the current draw of an electric motor.

The most important loads in refrigeration and conditioned air systems are electric motors. The technician must know how to correctly diagnose the condition of each type of electric motor that is used in the industry, including open-type motors used to rotate fan motors, pumps, open-type compressors, and other components that need rotating power; and enclosed-type motors, which are used in hermetic and semi-hermetic compressors. Electric motors will generally fail in three areas: windings, starting components, and bearings. The condition of the windings of an electric motor can easily be determined by checking the resistance of each winding. Single-phase electric motors have a start and run winding, as shown in (Figure 15.1), while three-phase motors have three or more sets of run windings, as shown in (Figure 15.2). Starting components like centrifugal switches used in open-type motors, starting relays used in hermetic compressors and some open-type motors, and capacitors must be checked if the motor is having difficulty starting. The bearings of the motor reduce friction and insure free movement of the motor rotor. Bearings of open-type motors can be diagnosed by turning the shaft manually; if the shaft is easily turned, the bearings are probably okay. The bearings of open and enclosed electric motors can also be checked with a clamp-on ammeter by determining the current draw of the motor when it is operating and comparing it to the listed full load amps of the motor. Once the technician has determined that an electric motor is faulty, then the correct replacement must be selected.

Contactors and relays are used in most heating, cooling, and refrigeration equipment for the operation of loads in the system such as motors, compressors, heaters, pumps, and so on. Contactors and relays are similar in their operation because both open and close a set or sets

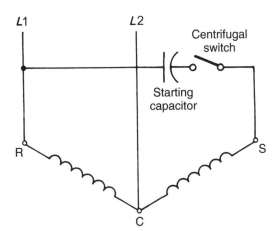

Figure 15.1 Single-phase electric motor with centrifugal switch.

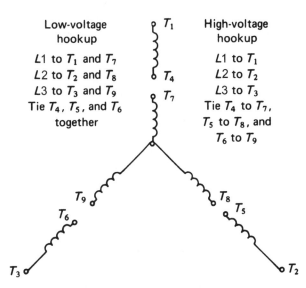

Figure 15.2 Three-phase star winding of a three-phase motor.

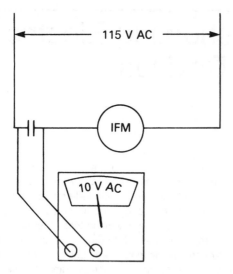

Figure 15.3 Voltage test of a set of contacts: 10 volts AC are lost across the contacts.

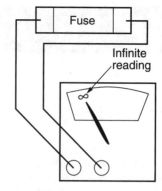

Figure 15.4 Resistance check of bad fuse.

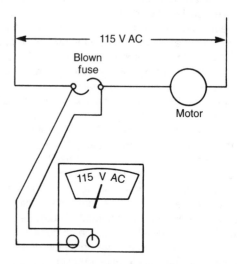

Figure 15.5 Voltage check of bad fuse.

of contacts when their coil is energized. The contactor is larger and capable of carrying more current than the relay. The three areas of problems encountered with contactors and relays are faulty contacts, coils, and mechanical linkages. The same procedure can be used to check either contactors or relays. The contacts of a contactor or relay can be checked visually if the contacts can be seen. The technician can also make a resistance or voltage check of the contacts to determine their condition. This test must be done with the contactor closed. When making a voltage check, the power source must be supplied to the load through the contacts; if any voltage is read across the contacts, that is what is being lost through the contacts, as shown in (Figure 15.3). The resistance check is performed with the power source disconnected from the contacts. This check is normally performed on contacts that cannot be visually inspected. The resistance reading across a good set of contacts will be 0 ohm. The condition of the coil or solenoid of a contactor or relay can be determined by performing a resistance check of the coil. The coil should have a measurable resistance; this resistance could vary a great deal depending upon the voltage of the coil. The mechanical linkage of a contactor or relay seldom causes problems, but occasionally could stick open or closed, contacts could break away from the linkage, or wear could cause drag due to friction. The technician must be able to diagnose these conditions and replace contactors and relays in HVAC equipment and control circuits.

Most major loads used in the heating, cooling, and refrigeration industry have some type of overload protection. The high cost of these system components makes it necessary to protect them. Overloads can cause problems in the operation of loads when they malfunction, such as allowing loads to operate in an overloaded condition or when their contacts open permanently. The simplest type of overload is a fuse. It can be checked with an ohmmeter, as shown in (Figure 15.4). A bad fuse will read infinity while a good fuse will read 0 ohm. On occasion, fuses will read a measurable resistance, which indicates that the element is only partially blown. In this case, the fuse must be replaced. Fuses can also be checked with

a voltmeter, as shown in (Figure 15.5). The circuit should be checked in the same way as fuses except sometimes circuit breakers must also be checked to determine the accuracy of their trip current. When checking pilot duty overloads, both the sensing element and the control contacts must be checked. The continuity of the element and contacts must be checked to determine if the overload is in good condition. However, before condemning an overload, the technician must make certain that the overload is faulty and an overload has not occurred. The technician should use care when checking compressor internal overloads because the internal overload may require an extended period of time to reset after an overload has occurred. When checking the condition of overloads, make certain that the overload is doing what it is supposed to do and at the correct current or temperature.

Thermostats play an important part in the correct operation of heating, cooling, and refrigeration systems. Line voltage thermostats are easy to troubleshoot because of their simplicity, while the low-voltage thermostat has many functions and accessibility of terminals is more difficult. There are basically two problems that occur with thermostats: first, the thermostat does not open or close at a temperature close enough to keep the customer satisfied, and second, it does not open or close at all. The line voltage thermostat can easily be checked by checking the continuity of the contacts. The technician must make certain to know what the thermostat *should* be doing at each stage in the operating cycle.

NOTE: T/S indicates room thermostat.

	T/S jumpered: system won't work.	T/S jumpered: system works.	Room temp. overshoots	Room temp. doesn't reach setting; too cold.	System cycles too often; too warm.	System doesn't cycle often enough.	Room temp. swings excessively.	POSSIBLE CAUSES:
	•							T/S not at fault; check elsewhere.
			•					T/S wiring hole not plugged; drafts.
					•	•		T/S not exposed to circulating air.
			•	•				T/S not mounted level (mercury switch types).
			•	•				T/S not properly calibrated.
			•		•			T/S exposed to sun, source of heat.
		•				•		T/S contacts dirty.
		•		•				T/S set point too high.
			•					T/S set point too low.
		•		•				T/S damaged.
				•				T/S located too near cold air register.
		•						Break in T/S circuit.
			•	•		•	•	System sized improperly.

Figure 15.6 Troubleshooting chart for thermostat. *(Courtesy of Honeywell, Inc.)*

For example, if the thermostat should be closed and it is not, then it must be replaced. The low-voltage thermostat is harder to diagnose because of the additional functions and accessibility difficulties. The technician would have to check the low-voltage thermostat at a common junction point or remove the thermostat from the subbase and jump out the desired terminals directly on the thermostat. A troubleshooting chart for a typical low-voltage thermostat is shown in (Figure 15.6). The technician must make certain to become familiar with the operation of the thermostat before diagnosing problems. Thermostats are difficult to calibrate, and often it is more economical to change the thermostat than to attempt calibration.

Pressure switches are used on many HVAC systems. The most important fact that the technician should know when diagnosing the condition of a pressure switch is its function in the system. Once the function of the pressure switch is known, determine what the action of the pressure is and whether or not it is appropriate for the present condition.

Be especially careful when diagnosing heating controls used on fossil fuel systems because a flame is available that can cause extensive damage to the structure and equipment. Always make sure that the combustion chamber of fossil fuel heating appliances is in good condition because of hazards to the occupants of the structure. A troubleshooting chart for a system with a thermocouple is shown in (Figure 15.7). A troubleshooting chart for an intermittent pilot ignition system is shown in (Figure 15.8).

SAFETY SHUTOFF TROUBLESHOOTING CHART

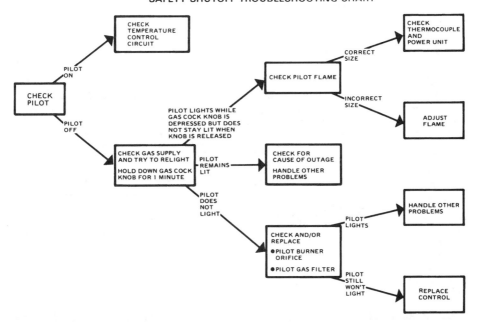

Figure 15.7 Troubleshooting chart for combination gas valve and thermocouple. *(Courtesy of Honeywell, Inc.)*

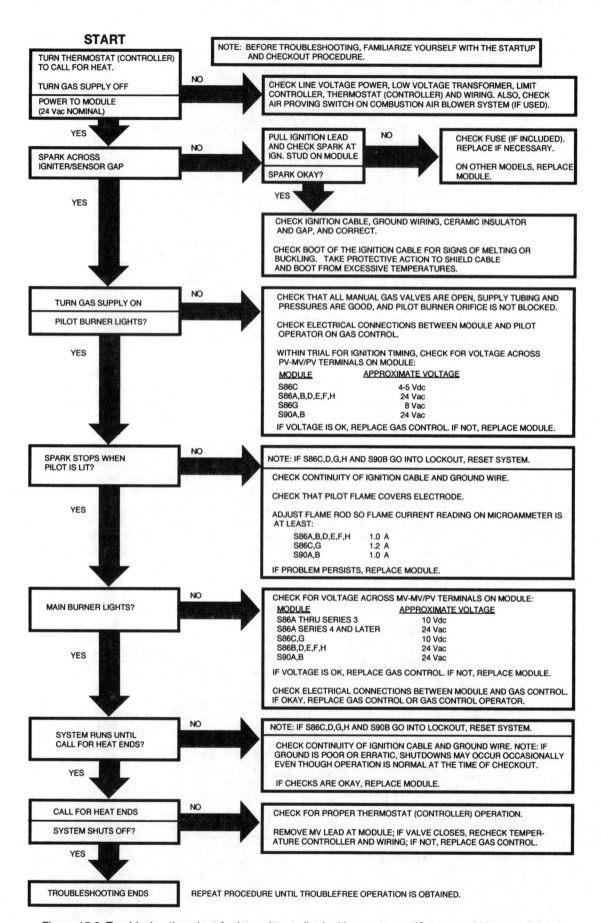

START

TURN THERMOSTAT (CONTROLLER) TO CALL FOR HEAT.

TURN GAS SUPPLY OFF

POWER TO MODULE (24 Vac NOMINAL)

NOTE: BEFORE TROUBLESHOOTING, FAMILIARIZE YOURSELF WITH THE STARTUP AND CHECKOUT PROCEDURE.

NO → CHECK LINE VOLTAGE POWER, LOW VOLTAGE TRANSFORMER, LIMIT CONTROLLER, THERMOSTAT (CONTROLLER) AND WIRING. ALSO, CHECK AIR PROVING SWITCH ON COMBUSTION AIR BLOWER SYSTEM (IF USED).

YES

SPARK ACROSS IGNITER/SENSOR GAP

NO → PULL IGNITION LEAD AND CHECK SPARK AT IGN. STUD ON MODULE

SPARK OKAY?

NO → CHECK FUSE (IF INCLUDED). REPLACE IF NECESSARY.

ON OTHER MODELS, REPLACE MODULE.

YES

YES

CHECK IGNITION CABLE, GROUND WIRING, CERAMIC INSULATOR AND GAP, AND CORRECT.

CHECK BOOT OF THE IGNITION CABLE FOR SIGNS OF MELTING OR BUCKLING. TAKE PROTECTIVE ACTION TO SHIELD CABLE AND BOOT FROM EXCESSIVE TEMPERATURES.

TURN GAS SUPPLY ON

PILOT BURNER LIGHTS?

NO → CHECK THAT ALL MANUAL GAS VALVES ARE OPEN, SUPPLY TUBING AND PRESSURES ARE GOOD, AND PILOT BURNER ORIFICE IS NOT BLOCKED.

CHECK ELECTRICAL CONNECTIONS BETWEEN MODULE AND PILOT OPERATOR ON GAS CONTROL.

WITHIN TRIAL FOR IGNITION TIMING, CHECK FOR VOLTAGE ACROSS PV-MV/PV TERMINALS ON MODULE:

MODULE	APPROXIMATE VOLTAGE
S86C	4-5 Vdc
S86A,B,D,E,F,H	24 Vac
S86G	8 Vac
S90A,B	24 Vac

IF VOLTAGE IS OK, REPLACE GAS CONTROL. IF NOT, REPLACE MODULE.

YES

SPARK STOPS WHEN PILOT IS LIT?

NO → NOTE: IF S86C,D,G,H AND S90B GO INTO LOCKOUT, RESET SYSTEM.

CHECK CONTINUITY OF IGNITION CABLE AND GROUND WIRE.

CHECK THAT PILOT FLAME COVERS ELECTRODE.

ADJUST FLAME ROD SO FLAME CURRENT READING ON MICROAMMETER IS AT LEAST:

S86A,B,D,E,F,H	1.0 A
S86C,G	1.2 A
S90A,B	1.0 A

IF PROBLEM PERSISTS, REPLACE MODULE.

YES

MAIN BURNER LIGHTS?

NO → CHECK FOR VOLTAGE ACROSS MV-MV/PV TERMINALS ON MODULE:

MODULE	APPROXIMATE VOLTAGE
S86A THRU SERIES 3	10 Vdc
S86A SERIES 4 AND LATER	24 Vac
S86C,G	10 Vdc
S86B,D,E,F,H	24 Vac
S90A,B	24 Vac

IF VOLTAGE IS OK, REPLACE GAS CONTROL. IF NOT, REPLACE MODULE.

CHECK ELECTRICAL CONNECTIONS BETWEEN MODULE AND GAS CONTROL. IF OKAY, REPLACE GAS CONTROL OR GAS CONTROL OPERATOR.

YES

SYSTEM RUNS UNTIL CALL FOR HEAT ENDS?

NO → NOTE: IF S86C,D,G,H AND S90B GO INTO LOCKOUT, RESET SYSTEM.

CHECK CONTINUITY OF IGNITION CABLE AND GROUND WIRE. NOTE: IF GROUND IS POOR OR ERRATIC, SHUTDOWNS MAY OCCUR OCCASIONALLY EVEN THOUGH OPERATION IS NORMAL AT THE TIME OF CHECKOUT.

IF CHECKS ARE OKAY, REPLACE MODULE.

YES

CALL FOR HEAT ENDS

SYSTEM SHUTS OFF?

NO → CHECK FOR PROPER THERMOSTAT (CONTROLLER) OPERATION.

REMOVE MV LEAD AT MODULE; IF VALVE CLOSES, RECHECK TEMPERATURE CONTROLLER AND WIRING; IF NOT, REPLACE GAS CONTROL.

YES

TROUBLESHOOTING ENDS

REPEAT PROCEDURE UNTIL TROUBLEFREE OPERATION IS OBTAINED.

Figure 15.8 Troubleshooting chart for intermittent pilot ignition systems. *(Courtesy of Honeywell, Inc.)*

A troubleshooting chart for a direct spark ignition system is shown in (Figure 15.9). On an oil burner system, the technician must check the primary control as well as the oil burner components. The technician should make sure that the gas or oil burner and safety controls are operating properly. When troubleshooting electric heating appliances, the only electrical device that has not been covered is the sequencer. The sequencer has an element that controls up to five sets of contacts; the element and contacts both must be checked. It is imperative to leave a heating system operating safely when the service call is completed.

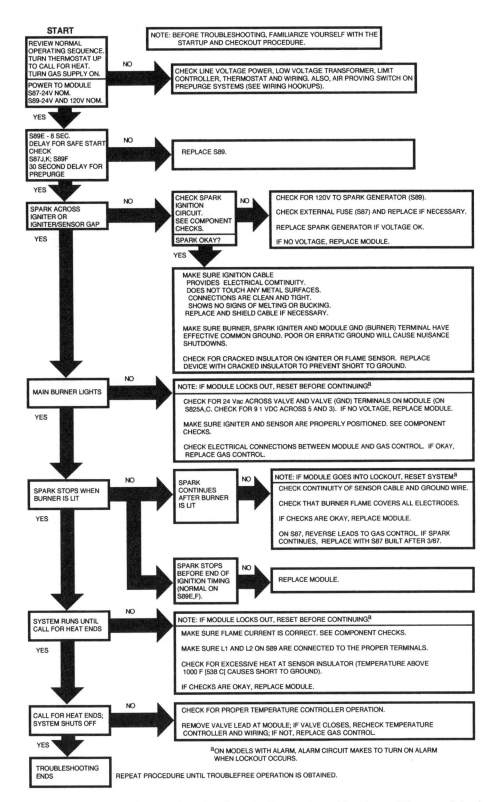

Figure 15.9 Troubleshooting chart for direct ignition system. *(Courtesy of Honeywell, Inc.)*

The transformer is an electrical device that is used to raise or lower the incoming voltage to the desired voltage via induction. The condition of the transformer can be determined by checking the continuity of the windings. However, there is always the possibility of a spot burnout in a transformer, which could allow the transformer to produce a voltage that does not meet system specifications.

The technician has the responsibility to become proficient at determining the condition of electrical components used in heating, cooling, and refrigeration systems. It is important that service personnel become technicians rather than parts changers.

Key Terms

Contactor

Ignition module

Motor

Pressure switch

Primary control

Relay

Thermostat

Transformer

REVIEW TEST

Name: _____   Date: _____   Grade: ___

Answer the following questions.

1. Why is the selection of a replacement motor important to the technician in the field?

2. What procedure could be used to determine the condition of the bearings in an open-type electric motor?

3. What procedure could be used to determine the condition of the bearings in a hermetic compressor motor?

4. How should a technician determine the condition of the windings in electric motors?

5. What are the methods that could be used to check the condition of the contacts of a contactor?

6. If the technician reads 50 volts across the contacts (L1 and T_1) of an energized contactor, what is the condition of the contacts?

7. What further checks should a technician make when replacing a 24-volt transformer?

8. What resistance reading would indicate a good and bad relay or contactor coil?

9. What are some of the problems caused by faulty mechanical linkages in contactors and relays?

10. How can a technician check the condition of the contacts in a sealed relay?

11. What is the simplest type of overload used in the HVAC industry today, and how is it checked?

12. What procedure should a technician use to determine the condition of a circuit breaker?

13. What is the difference between a noninductive and inductive load? What type of overload could be used for each?

14. What is the difference between a pilot duty overload and a line break overload?

15. What precautions should be taken when checking the internal load of a hermetic compressor?

16. What type of thermostat is used on each of the following applications?

   a. Operating control of window air conditioner

   b. Operating control of residential conditioned air system

   c. Fan switch used on warm air furnace

   d. Limit switch used on warm air furnace

17. What is the difference between checking a heating thermostat and checking a cooling thermostat?

18. What procedure is used to diagnose the condition of a low-voltage thermostat used on a residential heating and cooling system?

19. What two methods can be used to check a transformer?

20. What is the most important factor to know when checking a pressure switch?

21. What is the difference between checking the operation of a pressure switch used as an operating control and checking one used as a safety control?

22. How would a technician diagnose the condition of a sequencer that has four sets of contacts?

23. What procedure is used to check a standing pilot system with a combination gas valve?

24. What type of gas valve is used on an intermittent pilot system, and how is its condition checked?

25. How would a technician check the ignition module of an intermittent pilot system?

26. How would a technician check the ignition module of a direct ignition system?

27. What two methods are used to ignite the main burner of a direct ignition system, and how would a technician troubleshoot each?

28. How do a stack switch and a cad cell primary control sense that an oil burner has ignited?

29. If a cad cell failed to see an oil burner flame, what actions could be taken to correct the problem?

30. Why is it important to check the limit switches on a furnace on a preseason startup service call?

Troubleshooting Contactors and Relays

Name: _____	Date: _____	Grade: ___

Comments:

Objectives: Upon completion of this lab, you should be able to correctly determine the condition of contactors and relays.

Introduction: One of the most important tasks of the heating, refrigeration, and air-conditioning technician is trouble-shooting control systems. Contactors and relays are used to control loads in these control systems, and the technician must be able to correctly determine the condition of these components and replace them, if necessary.

Text References: Paragraphs 11.2 and 15.2

Tools and Materials: The following materials and equipment will be needed to complete this lab exercise.

Contactors

Relays

Contactor and relay kit

HVAC equipment

Electrical meters

Miscellaneous electrical supplies

Basic electrical handtools

Safety Precautions: Make certain that the electrical source is disconnected when making electrical connections. In addition:

*   Make sure all connections are tight.
*   Make sure no bare current-carrying conductors are touching metal surfaces except the grounding conductor.
*   Make sure the correct voltage is being supplied to the circuits.
*   Make sure body parts do not come in contact with live electrical conductors.
*   Keep hands and materials away from moving parts.

LABORATORY SEQUENCE (mark each box upon completion of task)

A. Troubleshooting Contactors and Relays

☐   1. Obtain the following contactors and relays from the supply room.
    a. 24-volt relay with 1 NO & 1 NC contacts
    b. 115-volt relay with 1 NO & 1 NC contacts
    c. 230-volt relay with 1 NO & 1 NC contacts
    d. 24-volt, 30-amp, 2-pole contactor
    e. 115-volt, 30-amp, 2-pole contactor
    f. 230-volt, 30-amp, 2-pole contactor
    g. 230-volt, 60-amp, 3-pole contactor

☐   2. Measure and record the resistance of the coils of the relays and contactors obtained in Step 1.

24-volt relay coil _____ ohms

115-volt relay coil _____ ohms

230-volt relay coil _____ ohms

24-volt contactor coil (2P) _____ ohms

115-volt contactor coil (2P) _____ ohms

230-volt contactor coil (2P) _____ ohms

230-volt contactor coil (3P) _____ ohms

☐   3. Connect the coils of the relays and contactors to the appropriate voltage. Measure and record the resistance of the normally open contacts when the component is de-energized and energized.

Relay NO contacts when relay is energized _____ ohms

Relay NO contacts when relay is de-energized _____ ohms

Relay NC contacts when relay is energized _____ ohms

Relay NC contacts when relay is de-energized _____ ohms

Contactor contacts when contactor is energized _____ ohms

Contactor contacts when contactor is de-energized _____ ohms

☐   4. Obtain a contactor and relay kit from your instructor. Troubleshoot and record the condition of the five contactors and five relays in the kit.

Relay or Contactor	Condition of Contactor or Relay
Relay #1	_____
Relay #2	_____
Relay #3	_____
Relay #4	_____
Relay #5	_____
Contactor #1	_____
Contactor #2	_____
Contactor #3	_____
Contactor #4	_____
Contactor #5	_____

☐   5. Have your instructor check your work.

☐   6. Return all contactors, relays, and contactor and relay kit to their appropriate location(s).

☐   7. Troubleshoot three contactors or relays on live equipment assigned by your instructor.

Relay or Contactor	Condition of Relay or Contactor
Indoor fan relay	_____
Contactor	_____
Relay or contactor	_____

☐   8. Have your instructor check your work.

MAINTENANCE OF WORK STATION AND TOOLS: Clean and return all tools to their proper location(s). Replace all equipment covers. Clean up the work area.

SUMMARY STATEMENT: Why is it important for the technician to be able to correctly troubleshoot and replace contactors and relays?

Questions

1. What is the approximate resistance of a 24-volt, 115-volt, and 230-volt relay coil?

2. What is the resistance of a set of NO contacts when a relay is de-energized?

3. What is the resistance of a set of NO contacts when a relay is energized?

4. What is likely to be the problem with a contactor if the contactor coil is good and is receiving the proper voltage but will not close?

5. How can a service technician determine the condition of the contacts of a contactor?

6. What other damage could a shorted contactor coil cause in a 24-volt control circuit?

7. What would be the likely problem with a relay if the contactor coil is receiving 24 volts but is not closing the contacts?

8. If a technician reads 80 volts across a set of contacts ($L1$–T_1) on a contactor, what is the problem?

9. What could be the results of mounting a contactor with a swinging armature in the wrong position?

10. How can the service technician determine the location of a specific relay in a control panel?

Troubleshooting Overloads

Name:	Date:	Grade:

Comments:

Objectives: Upon completion of this lab, you should be able to correctly determine the condition of overload devices.

Introduction: All major loads in a heating, refrigeration, and air-conditioning system will have some type of overload protection. It is the responsibility of the service technician to make certain that the overload protection is operating properly and, if not, to replace it with the correct replacement.

Text References: Paragraphs 11.3 and 15.3

Tools and Materials: The following materials and equipment will be needed to complete this lab exercise.
Line break overloads   Electrical meters
Pilot duty current-type overloads   Miscellaneous electrical supplies
Compressor with internal overload   Basic electrical handtools
HVAC equipment

Safety Precautions: Make certain that the electrical source is disconnected when making electrical connections. In addition:
- Make sure all connections are tight.
- Make sure no bare current-carrying conductors are touching metal surfaces except the grounding conductor.
- Make sure the correct voltage is being supplied to the circuits.
- Make sure body parts do not come in contact with live electrical conductors.
- Keep hands and materials away from moving parts.

LABORATORY SEQUENCE (mark each box upon completion of task)

A. Troubleshooting Overloads

☐ 1. Obtain the following overloads from the supply room.
   a. Line break current overload
   b. Pilot duty current overload
   c. Compressor with internal overload

☐ 2. Measure and record the resistance of the line break overload. _____ ohms

☐ 3. Measure and record the resistance of the control contacts and the controlling element of the pilot duty overload.

   Control contacts _____ ohms

   Controlling element _____ ohms

☐ 4. Measure and record the resistance of the compressor motor to determine the condition of the internal overload. (NOTE: The internal overload is electrically connected among common and start and run.)

Compressor Resistances

Common to start _____ ohms

Common to run _____ ohms

Start to run _____ ohms

☐ 5. Obtain an overload kit from your instructor. Troubleshoot and record the condition of the overloads in the kit.

Line break overload #1 _____

Line break overload #2 _____

Line break overload #3 _____

Pilot duty overload #1 _____

Pilot duty overload #2 _____

Pilot duty overload #3 _____

☐ 6. Have your instructor check your work.

☐ 7. Return all overloads, compressors, and the overload kit to their appropriate location(s).

☐ 8. Troubleshoot the three overloads on live equipment assigned by your instructor.

Overload	Condition of Overload
Line break overload	_____
Pilot duty overload	_____
Internal overload	_____

☐ 9. Have your instructor check your work.

MAINTENANCE OF WORK STATION AND TOOLS: Clean and return all tools to their proper location(s). Replace all equipment covers. Clean up the work area.

SUMMARY STATEMENT: Why are pilot duty overloads used on large loads instead of line break overloads?

Questions

1. Why is the trip-out current draw important in overload applications?

2. What is the simplest overload used in the HVAC industry?

3. What is the resistance of a good fuse?

4. What precautions should be taken when a technician suspects that an internal overload is open?

5. What procedure would a service technician use to correctly diagnose the condition of an internal overload in a hermetic compressor?

6. What is the difference between an internal overload and an internal thermostat?

7. If a service technician suspected that an overload was tripping at a lower than normal current draw, what could be done to check the overload?

8. How do electronic overloads sense the current of the load?

9. If a hermetic compressor reads 0 ohm between start and run, what is wrong with the compressor?

10. Why can't a fuse adequately protect a compressor motor?

LAB 15–3 Troubleshooting Thermostats

Name: _____	Date: _____	Grade: ____

Comments:

Objectives: Upon completion of this lab, you should be able to correctly troubleshoot line voltage and low-voltage thermostats.

Introduction: The main function of most refrigeration and conditioned air systems is to maintain a specific temperature of an object or space. The thermostat is the electrical device that controls the major loads in an HVAC system. The service technician must be able to troubleshoot line voltage and low-voltage thermostats in order to effectively service HVAC systems.

Text References: Paragraphs 12.2, 12.3, 12.4, and 15.4

Tools and Materials: The following materials and equipment will be needed to complete this lab exercise.
Line voltage thermostat
Single-stage cooling and heating low-voltage thermostat and subbase
Two-stage heating, two-stage cooling low-voltage thermostat and subbase
Thermostat kit
HVAC equipment
Electrical meters
Miscellaneous electrical supplies
Basic electrical handtools

Safety Precautions: Make certain that the electrical source is disconnected when making electrical connections. In addition:
- Make sure all connections are tight.
- Make sure no bare current-carrying conductors are touching metal surfaces except the grounding conductor.
- Make sure the correct voltage is being supplied to the circuit.
- Make sure body parts do not come in contact with live electrical conductors.
- Keep hands and materials away from moving parts.

LABORATORY SEQUENCE (mark each box upon completion of task)

A. Line Voltage Thermostats

☐ 1. Obtain line voltage heating and cooling thermostats from the supply room.

☐ 2. Using an ohmmeter, determine the action of the contacts of the line voltage cooling thermostat as the temperature setting is increased and decreased. What is the action of the thermostat contacts on a rise and fall in temperature?

☐ 3. Using an ohmmeter, determine the action of the contacts of the line voltage heating thermostat as the temperature setting is increased and decreased. What is the action of the thermostat contacts on a rise and fall in temperature?

☐ 4. Have your instructor check your work.

B. Low-Voltage Heating and Cooling Thermostat

☐   1. Obtain a heating and cooling low-voltage thermostat and subbase from the supply room.

☐   2. Connect thermostat wires from the R, W, Y, and G terminals of the low-voltage heating and cooling subbase.

☐   3. Route the thermostat wires through the opening in the thermostat subbase.

☐   4. Attach the thermostat to the subbase.

☐   5. Record the actions of the thermostat when the following selections are made. Check the continuity of the circuit in the thermostat using an ohmmeter.

     a. Move the fan switch from "Auto" to "On."

_____

     b. Fan switch in "Auto" position and system switch set to "Off."

_____

     c. Fan switch in "Auto" position, system switch set to "Cool" position, and thermostat set point decreased to a temperature that is lower than room temperature.

_____

     d. Fan switch in "Auto" position, system switch set to "Heat" position, and thermostat set point increased to a temperature that is higher than room temperature.

_____

☐   6. Have your instructor check your work.

C. Two-Stage Heating, Two-Stage Cooling Thermostat

☐   1. Obtain a two-stage heating, two-stage cooling low-voltage thermostat and subbase from the supply room.

☐   2. Connect thermostat wires from R, W1, W2, Y1, Y2, and G terminals of the two-stage heating, two-stage cooling subbase.

☐   3. Route the thermostat wires through the opening in the thermostat subbase.

☐   4. Attach the thermostat to the subbase.

☐   5. Record the actions of the thermostat when the following selections are made. Check the continuity of the thermostat using an ohmmeter.

     a. Move the fan switch from "Auto" to "On."

_____

     b. Fan switch in "Auto" position and system switch set to "Off."

_____

c. Fan switch in "Auto" position, system switch set to "Cool" position, and thermostat set at least 5°F below room temperature.

_____

d. Fan switch in "Auto" position, system switch set to "Heat" position, and thermostat set at least 5°F above room temperature.

_____

☐ 6. Have your instructor check your work.

☐ 7. Return all thermostats to their appropriate location.

D. Troubleshooting Thermostats

☐ 1. Obtain a thermostat kit from your instructor.

☐ 2. Troubleshoot and record the condition of the thermostats.

Line voltage thermostat #1 _____

Line voltage thermostat #2 _____

Line voltage thermostat #3 _____

Low-voltage thermostat #1 _____

Low-voltage thermostat #2 _____

Low-voltage thermostat #3 _____

☐ 3. Have your instructor check your work.

☐ 4. Return the thermostat kit to your instructor.

E. Troubleshooting Thermostats on Live Equipment

☐ 1. Troubleshoot the three thermostats on live equipment assigned by your instructor.

Thermostat	Condition of Thermostat
Window unit	_____
Gas heat/electric AC	_____
Heat pump	_____

☐ 2. Have your instructor check your work.

MAINTENANCE OF WORK STATION AND TOOLS: Clean and return all tools to their proper location(s). Replace all equipment covers. Clean up the work area.

SUMMARY STATEMENT: What is the difference between troubleshooting a line voltage thermostat and troubleshooting a low-voltage thermostat?

Questions

1. Name some common applications in which low-voltage thermostats are used.

2. Name some common applications in which line voltage thermostats are used.

3. Explain the action of a heating and cooling thermostat.

4. What is proper procedure for troubleshooting a two-stage heating, two-stage cooling thermostat?

5. How is the fan operated on a gas heat, electric air-conditioning system?

6. Why is a low-voltage thermostat more difficult to troubleshoot than a line voltage thermostat?

7. How would a service technician determine if a low-voltage thermostat is faulty?

8. How would a service technician jumper a low-voltage thermostat to determine if it is operating properly?

9. What is the best action to take if a thermostat has inaccurate cut-in and cut-out temperatures?

10. What would happen if the capillary tube connecting the thermostat to the bulb was broken?

Troubleshooting Pressure Switches

Name: _____	Date: _____ Grade: ___
Comments:	

Objectives: Upon completion of this lab, you should be able to correctly troubleshoot line voltage and low-voltage thermostats on equipment in the shop.

Introduction: Pressure switches are used as safety controls and operating controls in the control systems of refrigeration, heating, and air-conditioning systems. The service technician must know the function of the pressure switch in the control system before attempting to troubleshoot it. When troubleshooting pressure switches, the technician must know at what point the pressure switch should be opening or closing and at what point it actually is opening or closing. Pressure switches are used to protect the refrigeration system from operating at pressures that are unsafe.

Text References: Paragraphs 12.5 and 15.5

Tools and Materials: The following materials and equipment will be needed to complete this lab exercise.

Low-pressure switches	Gauge manifold set
High-pressure switches	Basic handtools
Dual-pressure switches	Electrical meters
Kit of pressure switches	Miscellaneous electrical supplies
HVAC equipment	Basic electrical handtools

Safety Precautions: Make certain that the electrical source is disconnected when making electrical connections. In addition:
- Make sure all connections are tight.
- Make sure no bare current-carrying conductors are touching metal surfaces except the grounding conductor.
- Make sure the correct voltage is being supplied to the circuits.
- Make sure body parts do not come in contact with live electrical conductors.
- Keep hands and materials away from moving parts.

LABORATORY SEQUENCE (mark each box upon completion of task)

A. Determining the Condition of Pressure Switches

☐   1. Obtain low-pressure, high-pressure, and dual-pressure switches from the supply room.

☐   2. Set the low-pressure switch (opens on a decrease in pressure) to cut in at 30 psig and cut out at 15 psig.

☐   3. Attach a pressure source to the pressure switch and determine the cut-in and cut-out pressures.

       Cut-in pressure _____ psig

       Cut-out pressure _____ psig

☐   4. Set the high-pressure switch (opens on a rise in pressure) to cut out at 100 psig and cut in at 75 psig.

☐   5. Attach a pressure source to the pressure switch and determine the cut-in and cut-out pressures.

       Cut-in pressure _____ psig

       Cut-out pressure _____ psig

☐ 6. Set the low-pressure switch of the dual-pressure switch to cut out at 15 psig and cut in at 30 psig. Set the high-pressure switch of the dual-pressure switch to cut out at 150 psig with a fixed differential.

☐ 7. Attach a pressure source to both sides of the dual-pressure switch and determine the cut-in and cut-out pressures.

Low Pressure

Cut-in pressure _____ psig

Cut-out pressure _____ psig

High Pressure

Cut-in pressure _____ psig

Cut-out pressure _____ psig

☐ 8. Have your instructor check your work.

☐ 9. Return pressure switches to the tool room.

B. Troubleshooting Pressure Switches

☐ 1. Obtain a pressure switch kit from your instructor.

☐ 2. Troubleshoot the five pressure switches in the kit and record their condition.

Pressure Switch	Condition
#1	_____
#2	_____
#3	_____
#4	_____
#5	_____

☐ 3. Have your instructor check your work.

☐ 4. Return the pressure switches to your instructor.

C. Troubleshooting Pressure Switches on Live Equipment

☐ 1. Troubleshoot the three pressure switches on live equipment assigned by your instructor.

Unit	Condition of Pressure Switch
#1	_____
#2	_____
#3	_____

☐ 2. Have your instructor check your work.

MAINTENANCE OF WORK STATION AND TOOLS: Clean and return all tools to their proper location(s). Replace all equipment covers. Clean up the work area.

SUMMARY STATEMENT: What is the most important factor that a technician must consider when troubleshooting a pressure switch?

Questions

1. How could a pressure switch be used as an operating control to control temperature?

2. What is a dual-pressure switch?

3. Why must a technician know what the pressure is in a refrigeration system before troubleshooting a pressure switch?

4. What would be the switching action of a low-pressure switch used as a safety control?

5. Why is differential important when setting pressure switches?

6. Why are some pressure switches nonadjustable?

7. What would be the switching action of a low-pressure switch used as an operating control?

8. What would be the switching action of a high-pressure switch used to maintain a constant head pressure by stopping and starting a condenser fan motor?

9. What would be the approximate setting of a low-pressure switch used to maintain 45°F in a walk-in cooler?

10. What procedure would a technician use to troubleshoot pressure switches?

LAB 15–5 Troubleshooting Heating Controls

Name: _____	Date: _____	Grade: ___

Comments:

Objectives: Upon completion of this lab, you should be able to correctly troubleshoot electric, gas, and oil heating controls.

Introduction: The air conditioning of structures during the winter months requires a source of heat to maintain the desired temperature level. This heating source must be safely controlled while maintaining the desired temperature in the structure. Heating controls are designed to take care of both of these functions in the system. The service technician must be able to locate and correct problems in heating appliances while making certain that the safety controls are operating properly.

Text References: Paragraphs 13.9, Chapter 14, paragraphs 15.7, 15.8, and 15.9

Tools and Materials: The following materials and equipment will be needed to complete this lab exercise.

Electric furnace	Electrical meters
Gas furnace with standing pilot	Miscellaneous electrical supplies
Gas furnace with intermittent ignition	Basic electrical handtools
Gas furnace with direct ignition	Parts to repair furnaces
Oil furnace with stack switch	Troubleshooting charts
Oil furnace with cad cell primary control	

Safety Precautions: Make certain that the electrical source is disconnected when making electrical connections. Make certain that no gas leaks exist. Locate and make sure gas supply cutoff valve is operating correctly. Shut off the gas supply to the device when installing, modifying, or repairing it. Allow at least five minutes for any unburned gas to leave the area before beginning work. Remember that LP gas is heavier than air and does not vent upward. When working on an oil-fired furnace, make sure there are no leaks and no excess fuel oil exists around the furnace or combustion chamber. In addition:

- Make sure all electrical connections are tight.
- Make sure no bare current-carrying conductors are touching metal surfaces except the grounding conductor.
- Make sure the correct voltage is being supplied to the circuit or appliance.
- Make sure body parts do not come in contact with live electrical conductors.
- Keep hands and materials away from moving parts.
- Make sure no leaks are present in gas or oil lines.

LABORATORY SEQUENCE (mark each box upon completion of task)

A. Troubleshooting Electric Heating Controls

☐ 1. Obtain from your instructor an assignment of an electric furnace that is not operating properly.

☐ 2. Turn the power supply off and check the system wiring for any loose or broken electrical connections.

☐ 3. Turn the power supply on and check for line voltage at the circuit breaker or fuse block. If line voltage is not available, locate and repair the problem.

☐ 4. With line voltage available to the electric furnace, set the thermostat to a call for heat. There should be 24 volts available to the sequencer or relay controlling the resistance heaters. If 24 volts are not available, check the thermostat, transformer, and, as a last resort, check the thermostat wire connecting the furnace and thermostat. Repair or replace any faulty component(s).

☐ 5. If 24 volts are supplied to the electric furnace controls, the first-stage heating elements and the blower motor should come on together. The remaining heating elements should come on as the sequencer closes the contacts. Check the voltage being supplied to each element and the current draw of each element. If voltage is available to the heating element but there is no current draw, the heating element is probably broken or there is an open fuse link or limit switch. If no voltage is available to the heating element(s), check across the contacts of the sequencer or relay with a voltmeter; if voltage is read, the contacts are bad. Repeat the procedure for each heating element in the appliance. Replace any electrical device found to be faulty.

☐ 6. Increase the setting of the thermostat to determine if the electric heating appliance is operating properly.

☐ 7. Check the voltage to each element and the current draw of each element to make sure all elements and the blower are operating properly.

☐ 8. Decrease the setting of the thermostat. Make sure that all elements and the blower are turned off.

☐ 9. Have your instructor check your work.

☐ 10. Replace all covers on equipment and clean up the work area.

B. Troubleshooting Gas Heating Controls, Standing Pilot

☐ 1. Obtain from your instructor an assignment of a gas furnace with a standing pilot that is not operating properly.

☐ 2. Visually inspect the gas furnace to determine if there are any loose or broken electrical connections.

☐ 3. Use the troubleshooting chart in (Figure 15.21) (page 509) to proceed with your conclusions and repair.

☐ 4. Replace any electrical devices found to be faulty.

☐ 5. Increase the setting of the thermostat to a call for heat. Observe the operation of the gas furnace to determine if it is operating properly. Check all safety controls for proper operation.

☐ 6. Decrease the setting of the thermostat and make sure the main burner cuts off. Make sure the fan cools the combustion chamber before stopping.

☐ 7. Have your instructor check your work.

☐ 8. Replace all equipment covers and clean up the work area.

C. Troubleshooting Gas Heating Controls, Intermittent Ignition Pilot

☐ 1. Obtain from your instructor an assignment of a gas furnace with an intermittent ignition system.

☐ 2. Visually inspect the gas furnace to determine if any loose or broken electrical connections are present.

☐ 3. Use an appropriate troubleshooting chart like the one in (Figure 15.23) (page 512) to isolate the problem.

☐ 4. Replace any electrical devices found to be faulty.

5. Increase the setting of the thermostat to a call for heat. Observe the operation of the gas furnace to determine if it is operating properly. Check all safety controls for proper operation.

6. Decrease the setting of the thermostat and make sure the burners cut off. Make sure the fan cools the combustion chamber before stopping.

7. Have your instructor check your work.

8. Replace all equipment covers and clean up the work area.

D. Troubleshooting Gas Heating Controls, Direct Ignition

1. Obtain from your instructor an assignment of a gas furnace with a direct ignition system.

2. Visually inspect the gas furnace to determine if any loose or broken electrical connections are present.

3. Use the appropriate troubleshooting chart like the one in (Figure 15.25) (page 515) to isolate the problem.

4. Replace any electrical devices found to be faulty.

5. Increase the setting of the thermostat to a call for heat. Observe the operation of the gas furnace to determine if it is operating properly. Check all safety controls for proper operation.

6. Decrease the setting of the thermostat and make sure the burners cut off. Make sure the fan cools the combustion chamber before stopping.

7. Have your instructor check your work.

8. Replace all equipment covers and clean up the work area.

E. Troubleshooting Oil Heating Controls, Stack Switch

1. Obtain from your instructor an assignment of an oil-fired furnace with a stack switch primary control.

2. Visually inspect the oil furnace and the stack switch for loose and broken electrical connections.

3. To completely troubleshoot an oil burner installation, both the burner and ignition systems, as well as the primary control, must be checked for proper operation and condition.

In this troubleshooting section, only the electrical components are possible faults.

4. If the trouble does not seem to be in the burner or ignition systems, check all limit switches to make sure they are closed.

5. Reset the safety switch on the primary control.

6. Make sure that line voltage is available to the primary control.

7. Check the thermostat to make sure it is closed.

8. Set the thermostat to call for heat.

9. Put the contacts of the bimetal in step by pulling the drive shaft lever out 1/4 inch and releasing.

- [ ] 10. Troubleshoot the primary control.

- [ ] 11. Replace any electrical devices found to be faulty.

- [ ] 12. Increase the setting of the thermostat to a call for heat. Observe the operation of the oil furnace to determine if it is operating properly. Check all safety controls for proper operation.

- [ ] 13. Decrease the setting of the thermostat and make sure the oil burner cuts off. Make sure the fan cools the combustion chamber before stopping.

- [ ] 14. Have your instructor check your work.

- [ ] 15. Replace all equipment covers and clean up the work area.

F. Troubleshooting Oil Heating Controls, Cad Cell

- [ ] 1. Obtain from your instructor an assignment of an oil-fired furnace with a cad cell primary control.

- [ ] 2. Visually inspect the oil furnace, cad cell, and primary control for loose and broken electrical connections.

- [ ] 3. To completely troubleshoot an oil burner installation, both the burner and ignition system, as well as the primary control, must be checked for proper operation and condition.

In this troubleshooting section, only the electrical components are possible faults.

- [ ] 4. If the trouble does not seem to be in the burner or ignition system, check all limit switches to make sure they are closed.

- [ ] 5. Inspect the position and cleanliness of the cad cell.

- [ ] 6. Reset the safety switch on the primary control.

- [ ] 7. Make sure that line voltage is available to the primary control.

- [ ] 8. Set the thermostat to a call for heat.

- [ ] 9. Troubleshoot the cad cell primary control.

- [ ] 10. Replace any electrical device found to be faulty.

- [ ] 11. Increase the setting of the thermostat to a call for heat. Observe the operation of the oil furnace to determine if it is operating properly. Check all safety controls for proper operation.

- [ ] 12. Decrease the setting of the thermostat and make sure the oil burner cuts off. Make sure the fan cools the combustion chamber before stopping.

- [ ] 13. Have your instructor check your work.

- [ ] 14. Replace all equipment covers and clean up the work area.

MAINTENANCE OF WORK STATION AND TOOLS: Clean and return all tools to their proper location(s). Replace all equipment covers. Clean up the work area.

SUMMARY STATEMENT: Why is it important for the technician to check the operation of the safety controls on a fossil fuel installation?

Questions

1. Why are electric resistance heaters designed so they are not all energized at the same time?

2. Explain the operation of a sequencer used to control an electric furnace with four resistance heaters and a blower motor.

3. Why is the location of the cad cell important in an oil furnace using a cad cell primary control?

4. How many devices would have to be checked on a standing pilot ignition system, and what is the procedure for checking each?

5. What is a redundant gas valve, and how is it checked?

6. What two methods are used to prove ignition in an oil-fired furnace?

7. What could be the problem in an electric furnace if voltage is being supplied to the resistance heater but the heater is not producing heat? How would a technician determine this action?

8. What two methods are used to ignite a gas burner on a direct ignition system, and how is each checked?

9. Why are limit switches important to the safe operation of warm air furnaces?

10. What procedure would a technician use to troubleshoot direct ignition and intermittent ignition systems on gas furnaces?

CHAPTER 16 Residential Air-Conditioning Control Systems

Chapter Overview

The control system for conditioned air systems is designed to supervise the operation of the electrical loads of the equipment in order to maintain the desired temperature set by the occupant of the structure. Most residential conditioned air control systems are multifunctional supervising the operation of the blower motor, the heating mode of operation, and the cooling mode of operation. The control system is not only responsible for the operation of the equipment to meet the comfort needs of the structure but also for the safe operation of the equipment and the safety of the structure and occupants. Not too many years ago the basic conditioned air control system used in residences was simple in design and function. Technology has moved residential control systems far beyond simple because consumers are demanding more efficient systems, better temperature control, improved air quality, and more. Variable-speed motors are being used in many residential control systems improving the comfort in the structure and efficiency and improving operating cost. Zone control residential systems have become popular because they give a much better control in the entire structure. There are many other innovations that are being pushed by the advancements in technology in the HVAC industry. The lab exercises in this chapter will range from simple to complex residential control systems.

The basic configuration of air-conditioning and heating equipment is important because it will many times dictate what kind of control system is utilized to produce the best and most efficient operation of the conditioned air system. A residential packaged air-conditioning unit or heat pump is manufactured with all components housed in one unit with the thermostat mounted in the structure. The installation of a packaged unit in a residence is shown in (figure 16.1). Packaged units oftentimes use a fossil fuel for the heating source and are sometimes called gas packs. Electric resistance heaters can be added to a packaged unit to produce a heating source for the structure. Split air-conditioning systems are basically made up of two sections, a condensing unit and an evaporator with an air source which could be a fan coil unit or a furnace that is gas, oil, or electric with an evaporator mounted in the airflow as shown in (figure 16.2).

Basic residential control circuitry will depend on options that are selected by the initial builder or owner or the owner when replacement equipment is selected. The control circuitry will range from simple to complex depending on which type of equipment or system is selected and the cost of the system. Generally, the cost of residential systems span a broad range with the builder model equipment being much cheaper than the top of the line higher efficiency equipment that is usually put

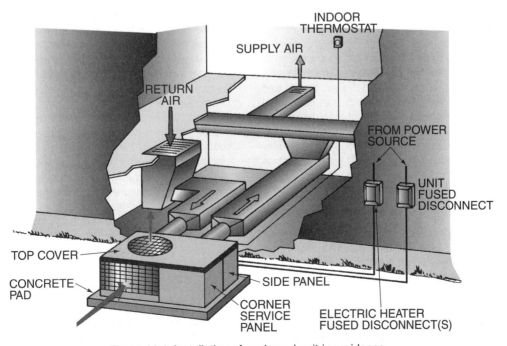

Figure 16.1 Installation of packaged unit in residence.

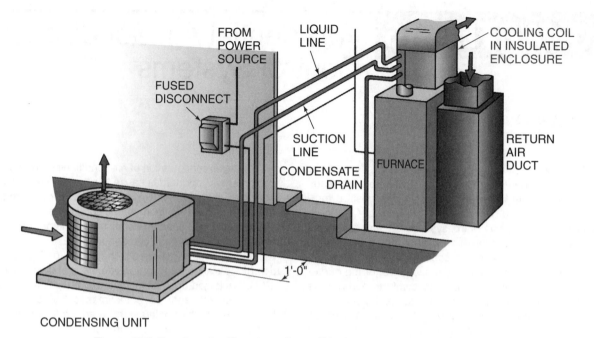

FROM POWER SOURCE

LIQUID LINE

COOLING COIL IN INSULATED ENCLOSURE

FUSED DISCONNECT

SUCTION LINE

CONDENSATE DRAIN

FURNACE

RETURN AIR DUCT

1'-0"

CONDENSING UNIT

Figure 16.2 Drawing of split system air-conditioning system with gas furnace.

in custom-built homes. A basic residential control circuit would have the simplest control system with the minimum number of controls that would be necessary for a system that supply adequate comfort and safety. For example, the schematic diagram of a small packaged unit is shown in (figure 16.3); this circuit has only the minimum necessary controls for comfort and safety. Top of the line and more efficient control systems generally operate with better comport and higher efficiency. No matter how basic or advanced the control system, certain functions are required. All residential conditioned air control systems must control the compressor, condenser fan motor, blower motor, and the heating source. Of course, if a system is using variable-speed motors, the control system will require a more sophisticated and complex control system with more controls. Any heating control system using fossil fuel, gas, or fuel oil as a heating source must supervise the ignition of the fuel as well as maintain safe operation of the equipment protecting the structure. Heat pump control circuits are more advanced than residential air conditioning circuit because air-to-air heat pumps must reverse the refrigeration cycle, defrost the outdoor coil on the heating cycle, and provide control of supplementary heat in the heating mode of operation. The basic function of the control system is to safely operate the loads in the system while maintaining the proper comfort level in the structure.

Over the past 25 years, the control systems used to control the operation of gas furnaces have advanced from control systems with standing pilots to those that ignite only when needed and then to a direct ignition system that lights the main burner. There are many standing pilots remaining in the industry but they are slowly but surely being replaced by the more efficient modern ignition systems. The standing pilot system uses a gas valve (see figure 14.4 in lab manual) that is equipped with a pilot solenoid that would keep the main gas valve

from opening if the pilot were not lit. The intermittent pilot system uses an electronic module (see figure 14.5 in lab manual) to supervise the ignition of the pilot (see figure 14.7 in lab manual) and when the pilot is proved the main gas valve will energize igniting the main burner. The direct ignition system uses an electronic module to supervise the prepurge cycle, the ignition of the main burner, postpurge cycle, blower motor operation, and diagnostics of gas operation. A gas furnace electronic module is shown in (figure 16.4).

Oil furnaces use stack switches (see Figure 14.9 in lab manual) or cad cell controls (see Figure 14.11 in lab manual) to supervise the operation of an oil-fired furnace. The cad cell is the most popular because its reaction to unsafe conditions is quicker than the stack switch. The cad cell (see Figure 14.10 in lab manual) sees the flame and sends a signal to the cad cell control that flame has be established. The cad cell actually sees the oil burner flame while the stack switch senses the flame by the temperature in the oil burner stack.

Almost all heat pumps presently being produced and installed in the industry are equipped with an electronic control module shown in (figure 16.5) that controls the electrical loads in the heat pump system. The schematic diagram for a heat pump with an electronic control module is shown in (figure 16.6). The heat pump electronic control module receives inputs from strategic points in the system that are relevant to the operation of the efficient operation of the heat pump. The module determines the mode of operation, when the system needs to be defrosted, if the system is operating at a unsafe condition, when supplementary heat is needed, and others functions.

The installation technician of air-conditioning and heating equipment must be able to connect the equipment to the thermostat in order to maintain the desired temperature in the structure. All manufacturers furnish wiring

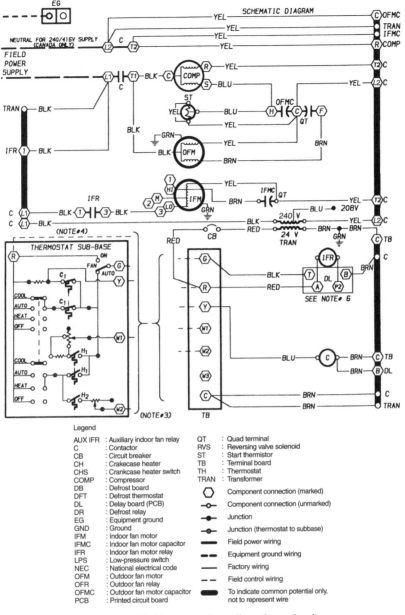

Figure 16.3 Diagram of small packaged unit.

Legend

AUX IFR	: Auxiliary indoor fan relay
C	: Contactor
CB	: Circuit breaker
CH	: Crakcase heater
CHS	: Crankcase heater switch
COMP	: Compressor
DB	: Defrost board
DFT	: Defrost thermostat
DL	: Delay board (PCB)
DR	: Defrost relay
EG	: Equipment ground
GND	: Ground
IFM	: Indoor fan motor
IFMC	: Indoor fan motor capacitor
IFR	: Indoor fan motor relay
LPS	: Low-pressure switch
NEC	: National electrical code
OFM	: Outdoor fan motor
OFR	: Outdoor fan relay
OFMC	: Outdoor fan motor capacitor
PCB	: Printed circuit board

QT	: Quad terminal
RVS	: Reversing valve solenoid
ST	: Start thermistor
TB	: Terminal board
TH	: Thermostat
TRAN	: Transformer

⬡ Component connection (marked)

─○─ Component connection (unmarked)

─●─ Junction

─⊘─ Junction (thermostat to subbase)

▬▬ Field power wiring

▬ ▬ Equipment ground wiring

─── Factory wiring

- - - Field control wiring

▬▬▬ To indicate common potential only, not to represent wire

Figure 16.4 Gas furnace electronic control module.

Figure 16.5 Electronic control module of heat pump.

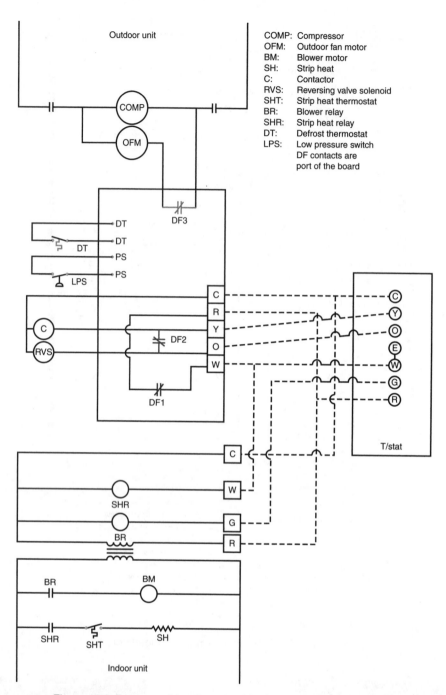

COMP: Compressor
OFM: Outdoor fan motor
BM: Blower motor
SH: Strip heat
C: Contactor
RVS: Reversing valve solenoid
SHT: Strip heat thermostat
BR: Blower relay
SHR: Strip heat relay
DT: Defrost thermostat
LPS: Low pressure switch
DF contacts are
port of the board

Figure 16.6 Diagram of heat pump with electronic control module.

diagrams with the equipment that shows proper connections of the line and low-voltage circuits. In some cases, the installation technician will make the line voltage connections unless otherwise specified by local codes, regulations, or specifications. It is important for the installation technician to make the proper electrical connection when installing electrical circuits on new installation. If line voltage connections are made by others, the installing technician should check and made sure that the electrician has made the proper connections. When equipment is being replaced, the HVAC technician will almost always make all of the electrical connections.

All heating, cooling, and refrigeration personnel must be familiar with the electrical connections that must be made on new and replacement installations. For example, sales personnel may be called on to assist customers with electrical system design. Engineers are required to design electrical systems both control and line voltage. Service technicians are required to install, maintain, and repair controls and line voltage systems.

Key Terms

Check, test, and start procedure
Condensing unit
Customer relations
Control wiring
Electric furnace

Factory-installed wiring
Fan-coil unit
Field wiring
Gas furnace
Heat pump

Oil furnace
Packaged unit
Power supply
Reversing valve
Split system

REVIEW TEST

Name: _____ Date: _____ Grade: ____

Complete the following multiple-choice questions by selecting the correct answer.

1. Which of the following is not a part of a condensing unit?
   a. condenser
   b. condenser fan motor
   c. compressor
   d. evaporator fan motor

2. An air-conditioning unit with all the components housed in one unit is known as a _____.
   a. split system
   b. packaged unit
   c. furnace
   d. condensing unit

3. What wiring is installed by the installation technician?
   a. code wiring
   b. strip wiring
   c. field wiring
   d. none of the above

4. What is the line and control voltage required by most residential condensing units?
   a. 115 volts/24 volts
   b. 115 volts/12 volts
   c. 230 volts/24 volts
   d. 460 volts/24 volts

5. The condenser fan motor and compressor of a small residential condensing unit are controlled by the _____.
   a. contactor
   b. circuit breaker
   c. condenser fan relay
   d. both a and c

6. The air source of a split air-condition system could be which of the following pieces of equipment?
   a. gas furnace
   b. oil furnace
   c. fan-coil unit
   d. all of the above

7. A packaged unit that is equipped with gas heat and electric air conditioning is sometimes called a _____.
   a. gas pack
   b. gas furnace with electric air-conditioning
   c. gas duct heater with air-conditioning
   d. none of the above

8. How many electrical power connections must be made when installing a condensing unit and a furnace?
   a. one
   b. two
   c. three
   d. none of the above

9. How many electrical power connections must be made when installing a gas pack?
   a. one
   b. two
   c. three
   d. none of the above

10. What device acts as a power source and a junction point for the control wiring connection of a split system with a gas furnace installation?
    a. circuit breaker
    b. terminal board
    c. split system package
    d. fan center

11. The condenser and compressor are wired in _____ in a small residential system.
    a. parallel
    b. series

12. What guide(s) should the installation technician use when sizing the power wire for an air-conditioning installation?
    a. National Electrical Code
    b. manufacturers installation instructions
    c. state and local codes
    d. all of the above

13. Which of the following letter designation supplies the reversing valve with 24 volts that energizes in the cooling mode of operation?
    a. R
    b. O
    c. W
    d. B

14. A hybrid heat uses _____ as the supplementary heat?
    a. electric resistance heaters
    b. reverse cycle
    c. a gas furnace
    d. nothing

15. Which of the following components is not a safety control in a condensing unit used for a residential air-conditioning system?
    a. high-pressure switch
    b. low-pressure switch
    c. contactor
    d. compressor internal overload

16. The outdoor coil of a heat pump is the _____ in the cooling mode of operation.
    a. evaporator
    b. condenser

17. A water-to-air heat pump uses water as the condensing medium in the _____ mode of operation.
    a. heating
    b. cooling

18. What component is responsible in a heater to change the flow of the refrigerant to change the modes of operation?
    a. Reversing plunger
    b. Flow reversal valve
    c. heating/cooling valve
    d. reversing valve

19. Which of the following items would need to be checked when a technician is completing a check, test, and start procedure on a new installation?
    a. line voltage to compressor
    b. suction pressure
    c. tightness of electrical connections
    d. all of the above

20. What is one of the most important elements of customer service?
    a. appearance
    b. communication skills
    c. clean-up work area
    d. all of the above

LAB 16–1 Identification of HVAC Equipment

Name: _____	Date: _____	Grade: ___

Comments:

Objectives: Upon completion of this lab, you should be able to:
- identify the airflow source for conditioned air systems
- identify the configuration of equipment for conditioned air systems
- identify the heating source for conditioned air systems

Introduction: Technicians in the HVAC industry will come in contact with all types and configurations of conditioned air equipment. The technician should be able to recognize the equipment type and configuration of any system on a new installation or a service call. It is important for the technician to determine the source of the air supply whether it be from a furnace or an air handler and the direction of flow, upflow, downflow, or horizontal. The two types of equipment are packaged units and split systems. The heating sources of conditioned air systems will be from a fossil fuel, gas or oil, electrical resistance heaters, or heat pumps.

Text Reference: Paragraph 16.1

Tools and Materials: The following materials and equipment will be needed to complete this lab exercise.
Manufacturer's installation instructions
Operating conditioned air systems:
Conditioning air system with fan coil unit
Conditioned air system with gas furnace
Packaged conditioned air system
Air-to-air heat pump with electrical resistance supplementary heat
Air-to-air heat pump with fossil fuel supplementary heat
Packaged heat pump
Water-to-air heat pump
Hand tools as necessary for the removal of equipment covers

Safety Precautions: When examining conditioned air equipment that is installed, make certain that the electrical power source is disconnected.

LABORATORY SEQUENCE (mark each box upon completion of task)

A. Identify airflow patterns in conditioned air systems

☐ 1. Your instructor will assign conditioned air systems on which to identify the air following airflow patterns and list the manufacturer and model number of the equipment:

a. upflow

Manufacturer _____

Model number _____

b. downflow

Manufacturer _____

Model number _____

c. horizontal

Manufacturer _____

Model number _____

B. Identify conditioned air equipment type

☐   1. Examine the conditioned air equipment in the laboratory and identify the following equipment types and list the manufacturer and model number of the equipment.

a. Electric air-conditioning packaged unit (AC only)

Manufacturer _____

Model number _____

b. Electric air-conditioning gas heating packaged unit

Manufacturer _____

Model number _____

c. Split air-conditioning system with air handler and condensing unit

Manufacturer _____

Model number (condensing unit) _____

(air handler) _____

d. Split air-conditioning system with fossil fuel furnace and condensing unit

Manufacturer _____

Model number (furnace) _____

(condensing unit) _____

e. Packaged air-to-air heat pump

Manufacturer _____

Model number _____

f. Water-to-air packaged heat pump

Manufacturer _____

Model number _____

g. Split system heat pump with electrical resistance heat as supplementary heat

Manufacturer _____

Model number (outdoor unit) _____

(indoor unit) _____

h. Split system heat pump with fossil fuel furnace as supplementary heat

Manufacturer _____

Model number (furnace) _____

(outdoor unit) _____

MAINTENANCE OF WORK STATION AND TOOLS: Clean and return all tools to their proper locations. Replace all covers used in this exercise.

SUMMARY STATEMENT: Describe the airflow patterns and the equipment types used in the HVAC industry.

Questions

1. In what HVAC application would a packaged air conditioning only unit be used?

2. What type of airflow configuration (upflow, downflow, horizontal) would be used in the following applications?
   a. Airflow appliance in the attic of a structure _____
   b. Airflow appliance in crawl space of a structure _____
   c. Airflow appliance in closet with duct work in attic _____
   d. Airflow appliance in closet with duct work in crawl space _____

3. What is the difference between an air-to-air heat pump and an air-to-water heat pump?

4. What is a gas pack?

5. What is the airflow of the conditioned air source of a conditioned air system using a fossil fuel furnace?

6. What is a fan-coil unit and where is it used?

7. On the heating cycle of a water-to-air heat pump, what is the condensing medium in the heating and cooling mode of operation?

8. Where is the evaporator placed in a conditioned air system using an oil-fired furnace as the heating and airflow source?

9. What heating sources could be used as the supplementary heat on heat pump systems?

10. What would be the location of the gas-fired furnace, the evaporator, and condensing unit in a split system installation?

Name: _____	Date: _____	Grade: ___

Comments:

Objectives: Upon completion of this lab, you should be able to correctly identify the electrical components of any residential conditioned air system.

Introduction: If a technician is to successfully install and troubleshoot conditioned air systems, he must be familiar with and be able to identify all electrical components used in the control system.

Text References: Chapter 16

Tools and Materials:
Packaged air conditioning unit
Split system air conditioning unit
Gas furnace
Split system heat pump

Safety Precautions: Make certain that the electrical source is disconnected when examining electrical control panels and components.

LABORATORY SEQUENCE (mark each box upon completion of task)

A. Locating Components in a Package Air Conditioning Unit

☐ 1. Your instructor will assign an air-cooled packaged unit on which to identify the following components:

  a. Condenser fan motor

  b. Evaporator fan motor

  c. Compressor motor

  d. Transformer

  e. Thermostat

  f. Indoor fan relay

  g. Contactor

  h. Capacitor

B. Locating Components in a Split System Air Conditioning System

☐ 1. Your instructor will assign you a split system air conditioning system on which to identify the following components:

    a. Condenser fan motor

    b. Evaporator fan motor

    c. Compressor motor

    d. Transformer

    e. Thermostat

    f. Indoor fan relay

    g. Contactor

    h. Capacitor

C. Locating Components on a Gas Furnace

☐ 1. Your instructor will assign you a gas furnace on which to identify the following components:

    a. Gas valve

    b. Furnace or ignition control module

    c. Hot surface igniter

    d. Pressure switch

    e. Limit switches

    f. Blower motor

    g. Induced draft motor

    h. Transformer

    i. Thermostat

    j. Door switch

    k. Light for status codes if applicable

    l. Control circuit fuse

D. Split System Heat Pump

☐ 1. Your instructor will assign you a split system heat pump with electric resistance heaters on which to identify the following components:

    a. Compressor

    b. Indoor fan motor

    c. Outdoor fan motor

    d. Reversing valve

e. Reversing valve solenoid

f. Heat pump control module

g. Supplementary heat

h. Supplementary heat overloads of limits

i. Contactor

j. Indoor fan relay

k. Transformer

l. Supplementary heat control plug if applicable

m. Supplementary heat line voltage plug if applicable

n. Supplementary heat contactor or sequencer

o. Thermostat

MAINTENANCE OF WORK STATION AND TOOLS: Clean and return all tools to their proper locations. Replace all covers on equipment used in this exercise.

SUMMARY STATEMENT: Why is it necessary for an HVAC technician to be able to identify the electrical components in a control system?

Questions

1. What is the purpose of a heat pump control module?

2. What is the purpose of a gas furnace control module?

3. What types of temperature limits are found on modern gas furnaces?

4. Why are temperature limits needed on supplementary heat elements?

5. Why are multi-speed indoor blowers used on most indoor blower motors?

6. Explain the difference between a thermostat used on a conditioned air system using a gas furnace as a heating source and a heat pump?

7. How does a reversing valve reverse the flow of refrigerant in a heat pump?

8. What is the purpose of an induced draft motor used on a gas furnace?

9. What control letter designation starts the compressor on the heating mode of operation on a heat pump and energizes the reversing valve on the cooling mode of operation?

10. What is the purpose of the pressure switch when used on a gas furnace?

LAB 16–3 Wiring an Air-conditioning Packaged Unit

Name: _____   Date: _____   Grade: ___

Comments:

Objectives: Upon completion of this lab, you should be able to:
- correctly size line voltage conductors for a packaged unit installation
- formulate a list of supplies that will be needed to make a packaged unit installation
- make the electrical connection for a packaged unit installation
- operate the packaged unit installed
- complete a check, test, and start checklist on installation

Introduction: Technicians in the HVAC industry will be called on to install conditioned air packaged units. The installation instructions usually furnished with new equipment along with local and state codes will be a guide for the installation of conditioned air equipment. The technician should be able to correctly size the line voltage conductors using the National Electrical Code or manufacturers installation instructions. A list of materials that will be needed to complete the installation should be compiled by the installation technician. The technician will be responsible to install and operate conditioned air equipment. Once equipment is installed and operating, the technician should complete a check test and start form.

Text Reference: Paragraphs 16.2, 16.3, 16.8 16.9, and 16.10.

Tools and Materials: The following materials and equipment will be needed to complete this lab exercise.
National Electrical Code
Manufacturer's installation instructions for unit assigned by instructor
Packaged air conditioning system
Low-voltage cooling thermostat
Electrical supplies (technicians list)
Tools needed to make installation

Safety Precautions: Make certain that the electrical source is disconnected when making electrical connections. In addition,
- Make sure all electrical connections are tight.
- Make sure no current-carrying conductors are touching metal surfaces except the grounding conductor.
- Make sure the correct voltage is being supplied to the equipment.
- Make sure all equipment covers are replaced.
- Make sure body parts do not come in contact with live electrical conductors.
- Keep hand and materials away from moving parts.

LABORATORY SEQUENCE (mark each box upon completion of task)

A. Install Package Air-Conditioning Unit

☐   1. See instructions for assignment of packaged air-conditioning unit.

☐   2. Read the section in the installation instructions on wiring the unit.

☐   3. Determine the correct wire size for the assigned unit. Take note of the distance between the packaged unit and the electrical supply panel. (NOTE: The type of insulation on the conductor plays an important part in its current-carrying ability.)

☐   4. Complete Data Sheet 16A.

DATA SHEET 16A

Package unit model number _____

Packaged unit serial number _____

Distance from power supply to unit _____

Size of wire to be installed _____

☐ 5. Complete Material List 16A including electrical supplies needed to complete installation.

MATERIALS LIST 16A

_____

_____

_____

_____

_____

☐ 6. Have your instructor check your wire sizing and material list for the assigned packaged unit installation.

☐ 7. Make the necessary electrical connections to the packaged unit assigned by the instructor.

☐ 8. Operate the unit and complete the CHECK, TEST, AND START LIST 16A

CHECK TEST AND START CHECKLIST 16A

Voltage being supplied to unit _____ volts

Amp draw of compressor _____ amps

Amp draw of outdoor fan motor _____ amps

Amp draw of indoor fan motor _____ amps

Supply air temperature _____ degrees F

Return air temperature _____ degrees F

CHECKLIST

All electrical connections tight	☐ YES	☐ NO
Electrical circuits properly labeled	☐ YES	☐ NO
Equipment properly grounded	☐ YES	☐ NO
All equipment level	☐ YES	☐ NO
All equipment covers in place and properly attached	☐ YES	☐ NO
Installation area clean	☐ YES	☐ NO
Homeowner instructed on operation of equipment	☐ YES	☐ NO

NOTE: The technician should also check the mechanical refrigeration cycle characteristics when performing check, test, and start procedures.

☐   9.   Have instructor check your installation.

☐   10.   Clean up the work area and return all tools and supplies to their correct locations.

MAINTENANCE OF WORK STATION AND TOOLS: Clean and return all tools to their proper locations.

SUMMARY STATEMENT: What procedure is used to size the conductor supplying an air-conditioning unit with more than one motor? Why should the technician perform a check, test, and start procedure on a new installation?

Questions

1. Explain the construction of a packaged unit?

2. Why are disconnects used in the installation of equipment?

3. What are two installation applications for packaged units?

4. How many electrical connections both must a technician make when installing a packaged unit?

5. What line voltage connections are required to complete the installation of a cooling-only packaged unit?

6. What low-voltage connections are required to complete the installation of a cooling-only packaged unit?

7. Why is it important for the technician to make sure that the unit is level?

8. Why should equipment be grounded?

9. Why should the installing technician check the amp draw of all electrical motors in a new installation.

10. What is the technician's responsibility to the homeowner at the completion of the installation?

<table>
<tr><td>Name: _____</td><td>Date: _____</td><td>Grade: ___</td></tr>
</table>

Comments:

Objectives: Upon completion of this lab, you should be able:
- Correctly size line voltage conductors for a condensing unit and fan coil unit installation.
- Formulate a list of supplies that will be needed to make a condensing unit with a fan coil unit installation.
- Make the electrical connection for a condensing unit with a fan coil unit.
- Operate the installed conditioned air system.
- Complete a check, test, and start checklist on the installation.

Introduction: Technicians will oftentimes have to install split air-conditioning systems with a fan coil unit that houses the fan used as the air source and the evaporator. The installation instructions usually furnished with new equipment along with local and state codes will be the guide for the installation of conditioned air equipment. The technician should be able to correctly size the line voltage conductors using the National Electrical Code or manufacturers installation instructions. A list of materials that will be needed to complete the installation should be compiled by the installation technician. The technician will be responsible to install and operate conditioned air equipment. Once equipment is installed and operating, the technician should complete a check test and start form.

Text References: Paragraphs 16.2, 16.4, 16.8, 16.19, and 16.10.

Tools and Materials: The following materials and equipment will be needed to complete this lab exercise.
National Electrical Code
Manufacturer's installation instructions for unit assigned by instructor
Air-conditioning condensing unit
Fan coil unit
Low-voltage cooling thermostat
Electrical supplies (technicians list)
Tools needed to make installation

Safety Precautions: Make certain that the electrical source is disconnected when making electrical connections. In addition,
- Make sure all electrical connections are tight.
- Make sure no current-carrying conductors are touching metal surfaces except the grounding conductor.
- Make sure the correct voltage is being supplied to the equipment.
- Make sure all equipment covers are replaced.
- Make sure body parts do not come in contact with live electrical conductors.
- Keep hand and materials away from moving parts.

LABORATORY SEQUENCE (mark each box upon completion of task)

A. Install Conditioned Air Split System (condensing unit and fan coil unit)

☐   1. See instructor for assignment of split conditioned air system including condensing unit and fan coil unit.

☐   2. Read the section in the installation instructions on wiring the unit.

☐   3. Determine the correct wire size for the assigned unit components (condensing unit and fan coil unit). Take note of the distance between the equipment and the electrical supply panel. (NOTE: The type of insulation on the conductor plays an important part in its current-carrying ability.)

☐   4. Complete Data Sheet 16B.

DATA SHEET 16B

Condensing unit model number _____

Condensing unit serial number _____

Distance from power supply to condensing unit _____

Size of wire to be installed for condensing unit _____

Fan coil unit model number _____

Fan coil unit serial number _____

Distance from power supply to unit _____

Size of wire to be installed for fan coil unit _____

☐   5. Complete Material List 16B including electrical supplies needed to complete installation.

MATERIALS LIST 16B

_____

_____

_____

_____

_____

_____

☐   6. Have your instructor check your wire sizing and material list for the assigned conditioned air system installation.

☐   7. Make the necessary electrical connections to the conditioned air system assigned by your instructor.

☐   8. Operate the conditioned air system and complete the CHECK, TEST, AND START LIST 16B.

CHECK TEST AND START CHECKLIST 16B

Voltage being supplied to condensing unit _____

Amp draw of compressor _____

Amp draw of condenser fan motor _____

Voltage being supplied to the fan coil unit _____

Amp draw of blower motor _____

Supply air temperature _____ degrees F

Return air temperature _____ degrees F

CHECKLIST

All electrical connections tight	☐ YES	☐ NO
Electrical circuits properly labeled	☐ YES	☐ NO
Equipment properly grounded	☐ YES	☐ NO
All equipment level	☐ YES	☐ NO
All equipment covers in place and properly attached	☐ YES	☐ NO
Installation area clean	☐ YES	☐ NO
Homeowner instructed on operation of system	☐ YES	☐ NO

NOTE: The technician should also check the mechanical refrigeration cycle characteristics when performing check, test, and start procedures.

☐ 9. Have instructor check your installation.

☐ 10. Clean up the work area and return all tools and supplies to their correct locations.

MAINTENANCE OF WORK STATION AND TOOLS: Clean and return all tools to their proper locations.

SUMMARY STATEMENT: What is the purpose of a disconnect switch on a condensing unit installation? Why should a technician check the amp draw of loads when a conditioned air system is installed?

Questions

1. What would be the application of using a fan coil unit?

2. Where would a split system conditioned air system be used?

3. How many power supply connections must a technician make when installing a conditioned air system using a fan coil unit and a condensing unit?

4. What controls wires are connected between the fan coil unit control connection and the thermostat?

5. What control wires are connected between the fan coil unit control connection and the condensing unit?

6. What type of thermostat unit would be used on a conditioned air-cooling-only installation of a condensing unit and a fan coil unit?

7. Why should the installation technician make certain that the unit is level?

8. Why should the installation technician check the supply and return air temperature upon completion of an installation?

9. What is the purpose of labeling the power supply circuits of a new installation?

10. Why is it important for an installation technician to communicate with the homeowner on new installation in an existing home?

Wiring a Split System Conditioned Air System with a Gas Furnace and Condensing Unit

Name: _____   Date: _____   Grade: ___

Comments:

Objectives: Upon completion of this lab, you should be able:
- Correctly size line voltage conductors for a condensing unit and gas furnace installation.
- Formulate a list of supplies that will be needed to make a condensing unit with a gas furnace installation.
- Make the electrical connection for a condensing unit with a gas furnace.
- Operate the installed conditioned air system.
- Complete a check, test, and start checklist on the installation.

Introduction: Technicians will oftentimes have to install split air-conditioning systems with a gas furnace that houses the fan used as the air source. The installation instructions usually furnished with new equipment along with local and state codes will be the guide for the installation of conditioned air equipment. The technician should be able to correctly size the line voltage conductors using the National Electrical Code or manufacturers installation instructions. A list of materials that will be needed to complete the installation should be compiled by the installation technician. The technician will be responsible to install and operate conditioned air equipment. Once equipment is installed and operating, the technician should complete a check test and start form.

Text References: Paragraphs 16.2, 16.4, 16.8, 16.9, and 16.10

Tools and Materials: The following materials and equipment will be needed to complete this lab exercise.
National Electrical Code
Manufacturer's installation instructions for unit assigned by instructor
Air-conditioning condensing unit
Gas furnace
Low-voltage heating/cooling thermostat
Electrical supplies (technicians list)
Tools needed to make installation

Safety Precautions: Make certain that the electrical source is disconnected when making electrical connections. In addition,
- Make sure all electrical connections are tight.
- Make sure no current-carrying conductors are touching metal surfaces except the grounding conductor.
- Make sure the correct voltage is being supplied to the equipment.
- Make sure all equipment covers are replaced.
- Make sure body parts do not come in contact with live electrical conductors.
- Keep hand and materials away from moving parts.

LABORATORY SEQUENCE (mark each box upon completion of task)

A. Install Conditioned Air Split System (condensing unit and gas furnace)

☐ 1. See instructor for assignment of split conditioned air system including condensing unit and gas furnace.

☐ 2. Read the section in the installation instructions on wiring the unit.

☐ 3. Determine the correct wire size for the assigned unit components (condensing unit and fan coil unit). Take note of the distance between the equipment and the electrical supply panel. (NOTE: The type of insulation on the conductor plays an important part in its current-carrying ability.)

☐ 4. Complete Data Sheet 16C.

DATA SHEET 16C

Condensing unit model number _____

Condensing unit serial number _____

Distance from power supply to condensing unit _____

Size of wire to be installed for condensing unit _____

Gas furnace model number _____

Gas furnace serial number _____

Distance from power supply to gas furnace _____

Size of wire to be installed for gas furnace _____

5. Complete Material List 16C including electrical supplies needed to complete installation.

MATERIALS LIST 16C

_____

_____

_____

_____

_____

_____

6. Have your instructor check your wire sizing and material list for the assigned conditioned air system installation.

7. Make the necessary electrical connections to the conditioned air system assigned by your instructor.

8. Operate the conditioned air system and complete the CHECK, TEST, AND START LIST 16C.

CHECK TEST AND START CHECKLIST 16C

Voltage being supplied to condensing unit _____

Amp draw of compressor _____

Amp draw of condenser fan motor _____

Voltage being supplied to the fan coil unit _____

Amp draw of blower motor _____

Supply air temperature _____ degrees F

Return air temperature _____ degrees F

CHECKLIST

All electrical connections tight	☐ YES	☐ NO
Electrical circuits properly labeled	☐ YES	☐ NO
Equipment properly grounded	☐ YES	☐ NO
All equipment level	☐ YES	☐ NO
All equipment covers in place and properly attached	☐ YES	☐ NO
Installation area clean	☐ YES	☐ NO
Homeowner instructed on operation of system	☐ YES	☐ NO

NOTE: The technician should also check the mechanical refrigeration cycle characteristics and gas heating operation when performing check, test, and start procedures.

☐ 9. Have the instructor check your installation.

☐ 10. Clean up the work area and return all tools and supplies to their correct locations

MAINTENANCE OF WORK STATION AND TOOLS: Clean and return all tools to their proper locations.

SUMMARY STATEMENT: What is the heating source of the conditioned air system installed in this lab exercise? Where is the evaporator located in this application?

Questions

1. Where is the connection point for the controls wiring from the thermostat and condensing unit?

2. What supervises the operation of the gas furnace in modern gas furnaces?

3. What controls the blower motor in modern gas furnaces?

4. What are the electrical components of a basic condensing unit?

5. What do the following letter designations represent in a split system utilizing a gas furnace as an air source and a condensing unit: "R," "G," "W," and "Y"?

6. What connection on a gas furnace terminal board would be supplied to the condensing unit?

7. What controls the condenser fan motor on a basic conditioned air-condensing unit?

8. What controls the blower motor on a gas-fired furnace?

9. How many power connections are required in the installation of a gas furnace and a condensing unit?

10. What is the control voltage of a conditioned air split system utilizing a gas furnace as the air source?

Wiring a Split System Heat Pump

Name: _____   Date: _____   Grade: ____

Comments:

Objectives: Upon completion of this lab, you should be able:
- Correctly size line voltage conductors for the outdoor unit, indoor unit, and supplementary electrical resistance heat of a split system heat pump.
- Formulate a list of supplies that will be needed to install the outdoor unit, indoor unit, and supplementary electrical resistance heat of a split system heat pump.
- Make the electrical connection for the outdoor unit, indoor unit, and supplementary electrical resistance heat of a split system heat pump.
- Operate the installed conditioned air system.
- Complete a check, test, and start checklist on the installation.

Introduction: Technicians will oftentimes have to install split system heat pumps along with some type of supplementary heat. The installation instructions usually furnished with new equipment along with local and state codes will be the guide for the installation of conditioned air equipment. The technician should be able to correctly size the line voltage conductors using the National Electrical Code or manufacturers installation instructions. A list of materials that will be needed to complete the installation should be compiled by the installation technician. The technician will be responsible to install and operate conditioned air equipment. Once equipment is installed and operating, the technician should complete a check test and start form.

Text References: Paragraphs 16.2, 16.5, 16.8, 16.9, and 16.10

Tools and Materials: The following materials and equipment will be needed to complete this lab exercise.
    National Electrical Code
    Manufacturer's installation instructions for unit assigned by instructor
    Outdoor unit of split system heat pump
    Indoor unit of split system heat pump
    Resistance heat for heat pump
    Low-voltage heat pump thermostat
    Electrical supplies (technicians list)
    Tools needed to make installation

Safety Precautions: Make certain that the electrical source is disconnected when making electrical connections. In addition,
- Make sure all electrical connections are tight.
- Make sure no current-carrying conductors are touching metal surfaces except the grounding conductor.
- Make sure the correct voltage is being supplied to the equipment.
- Make sure all equipment covers are replaced.
- Make sure body parts do not come in contact with live electrical conductors.
- Keep hand and materials away from moving parts.

LABORATORY SEQUENCE (mark each box upon completion of task)

A. Install Split System Heat Pump with Supplementary Heat

☐   1. See instructor for assignment of split conditioned air system including condensing unit and gas furnace.

☐   2. Read the section in the installation instructions on wiring the unit.

☐   3. Determine the correct wire size for the assigned unit components (outdoor unit and indoor unit). Take note of the distance between the equipment and the electrical supply panel. (NOTE: The type of insulation on the conductor plays an important part in its current-carrying ability.)

☐   4. Complete Data Sheet 16D.

DATA SHEET 16D

Outdoor unit model number _____

Outdoor unit serial number _____

Distance from power supply to condensing unit _____

Size of wire to be installed for condensing unit _____

Indoor unit model number _____

Resistance heat model number _____

Distance from power supply to indoor unit _____

Size of wire to be installed for indoor unit with supplementary heat _____

☐   5. Complete Material List 16D including electrical supplies needed to complete installation.

MATERIALS LIST 16D

_____

_____

_____

_____

_____

_____

☐   6. Have your instructor check your wire sizing and material list for the assigned conditioned air system installation.

☐   7. Make the necessary electrical connections to the conditioned air system assigned by your instructor.

☐   8. Operate the conditioned air system and complete the CHECK, TEST, AND START LIST 16D.

CHECK TEST AND START CHECKLIST 16D

Voltage being supplied to outdoor unit _____

Amp draw of compressor _____

Amp draw of outdoor fan motor _____

Voltage being supplied to the indoor unit _____

Amp draw of blower motor _____

Amp draw of resistance heaters _____

Supply air temperature (w/o supplementary heat) _____ degrees F

Supply air temperature (with supplementary heat) _____ degrees F

Return air temperature _____ degrees F

CHECKLIST

All electrical connections tight	☐ YES	☐ NO	
Electrical circuits properly labeled	☐ YES	☐ NO	
Equipment properly grounded	☐ YES	☐ NO	
All equipment level	☐ YES	☐ NO	
All equipment covers in place and properly attached	☐ YES	☐ NO	
Installation area clean	☐ YES	☐ NO	
Homeowner instructed on operation of system	☐ YES	☐ NO	

NOTE: The technician should also check the mechanical refrigeration cycle characteristics and gas heating operation when performing check, test, and start procedures.

☐　9. Have instructor check your installation.

☐　10. Clean up the work area and return all tools and supplies to their correct locations

MAINTENANCE OF WORK STATION AND TOOLS: Clean and return all tools to their proper locations.

SUMMARY STATEMENT: Why is supplementary heat necessary on an air-to-air heat pump? Why type of low-voltage thermostat is used on the heat pump in this lab exercise?

Questions

1. Why is supplementary heat required on most air-to-air heat pumps?

2. What is the difference between the mechanical refrigeration cycle of a heat pump and a regular air conditioner?

3. What are the common letter designations of most heat pumps and what mode of operation are related to these letter designations?

4. What two sources of supplementary heat could be used by modern heat pumps?

5. What is the purpose of the reversing valve in the mechanical refrigeration cycle of a heat pump?

6. What supervises the operation of modern heat pump?

7. What line voltage electrical connections must be made when installing an air-to-air split system heat pump using electrical resistance heater as the supplementary heat?

8. What control voltage connections must be made when installing an air-to-air split system heat pump using electrical resistance heaters as the supplementary heat?

9. How are electrical resistance heaters mounted in most indoor blower units in a split system heat pump installation?

10. Where is the indoor coil positioned in a split system heat pump installation when using a gas furnace as the supplementary heat?

CHAPTER 17 Commercial and Industrial Air-conditioning Control Systems

Chapter Overview

Control systems used in commercial and industrial heating and air-conditioning systems supervise the operation of a wide range of equipment from a 25-ton light commercial air-conditioning systems in small commercial buildings to a 1000-ton system that supplies comfort cooling and heating, cooling or heating mediums that are used in industrial processes, load management, and other functions. In the light commercial sector of the industry, control systems in many cases could be much like the residential units that have been discussed in a previous chapter in this lab manual. In light commercial and industrial control systems, the loads being controlled are larger requiring larger electrical devices to control them because of the higher current. With the increased cost of larger and more expensive equipment, the necessity of safely protecting larger loads is of paramount importance and many additional safety controls are used in these control systems. Larger commercial and industrial structures generally use a control system that is designed specifically for the structure, which could include 1) the control of the airflow into each part of the structure, 2) the comfort conditions in the structure, 3) load management of the structure, and 4) industrial processes that are required by the occupant.

The selection of the mechanical systems to be used in commercial and industrial structures in most cases is made by architects and mechanical engineers in the initial planning phase of the complex. The architect must examine a multitude of options before selecting the equipment to be used in the structure. Energy cost will play an important part in the selection of the equipment due to the importance of the economic element in the operation of the complex. The capacity of the heating and cooling load of the structure must be calculated, the selection of the medium used to condition the structure must be selected, the type of control system that is to be used, and the location of the equipment are factors that must be determined by the architect or engineer. Once these decisions have been made, the architect or engineer will have to produce a set of blueprint and specification for the structure. The general contractor will supervise the building of the structure. Mechanical contractors will be responsible for the installation of the mechanical systems. The HVAC technician will be responsible for the comfort cooling system to be used in the structure as well as other mechanical systems used for manufacturing processes such as chilled or hot water and steam.

The types of equipment used in the commercial and industrial sector of the HVAC industry are of a wide variety and usually larger than 25 tons capacity. Many commercial and industrial applications use commercial and industrial condensing units with large fan coil units to supply conditioned air to the structure. The commercial and industrial condensing unit houses a compressor, the condenser, condenser fan motors, and the controls necessary to safely supervise the operation of the unit and is shown in (Figure 17.1). This type of large condensing unit will utilize a large commercial and industrial fan coil unit to supply conditioned air and is shown in (Figure 17.2). A multizone rooftop packaged unit would usually be positioned on a roof and supply air to zones within a structure and is shown in (Figure 17.3). Packaged unit could also be place on ground level and supply air to a structure. Chilled and hot water systems are popular in commercial and industrial buildings and utilize a hot water boiler shown in (Figure 17.4)

Figure 17.1 Commercial and industrial air-conditioning condensing unit.

Figure 17.2 Commercial fan coil unit.

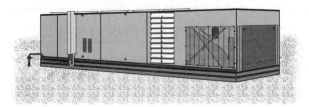

Figure 17.3 Multi-zone roof-top packaged unit.

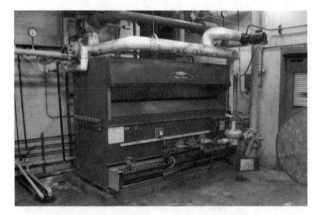

Figure 17.4 Hot water boiler.

Figure 17.5 Water chiller.

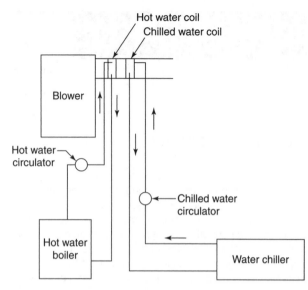

Figure 17.6 Piping of hot water and chilled water system.

to supply the hot water to the system and a water chiller shown in (Figure 17.5) to supply chilled water to the system. The chilled water supplied by the water chiller and hot water supplied by the hot water boiler is routed to a coil which air is passed over to supply what is needed in a particular zone this could be hot or chilled water. A drawing of this type of system is shown in (Figure 17.6). This by no means covers all of the types of systems used in commercial and industrial applications but only is an overview.

All types and designs of control systems are used in commercial and industrial conditioned air systems. The control systems used in equipment and systems above 25 tons capacity cannot afford to be as simple and use as few components as the smaller residential systems. This type of control system must ensure that the electric loads are operating in a safe manner because of their greater cost. The control of commercial and industrial equipment and control systems is generally more complex than the smaller systems because of its size and operation.

Most commercial and industrial systems use some method of capacity control because of the variations in the load in the conditioned areas. When capacity control is used on large air-conditioning systems, it usually involves some means of changing the compressor capacity to meet the load of the conditioned space. Compressor motors generally use an across-the-line starting, which energizes the motor with the closing of one contactor or part-winding motor hookup which brings on onset of winding approximately 1 to 3 seconds later than the first to increase the starting torque and to maintain a better load balance in a circuit. Water chiller controls systems must have some method to interlock the cool water pump with the operation of the chiller and if equipped with a water-cooled condenser must interlock the condenser water pump with the operation of the compressor. Blower motor control in commercial and industrial systems is generally accomplished through the mail HVAC control. With the advancement of electronic motor speed control, many air systems utilize a variable air volume system, which varies the airflow to any given zone. There are many types of control systems that are used in the commercial and industrial sector of the industry that are special systems designed for a specific complex and will be similar and different in many ways.

The air-cooled packaged unit with a remote condenser is a freestanding unit that houses the air supply source, compressor, and necessary controls to safely operate the system while the condenser is mounted remotely in a place where the heat rejected is unobjectionable. The air-cooled packaged unit is a freestanding unit that houses the air supply source, the compressor, condenser, and necessary controls to safely operate the system. A drawing of an air-cooled packaged unit is shown in (Figure 17.7). The water-cooled packaged unit has a water-cooled condenser with water being used as the condensing medium.

Most commercial and industrial control systems use a control loop to gather the information that is the input signal to some type of controller and the controller out puts signals to the controlled device that is used to

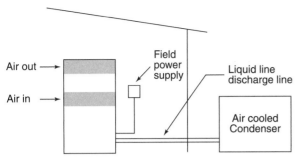

Figure 17.7 Drawing of air-cooled packaged unit.

maintain the temperature in a structure or zone. A control loop is shown in (Figure 17.8). At present, there are two controls systems that are being used in the HVAC industry and are the direct digital control (electronic) and pneumatic (air) control systems. The pneumatic control system uses air to transmit the signal from the sensor to the controller and on to the controlled device. (Figure 17.9) shows the diagram of a pneumatic control system. With the technological

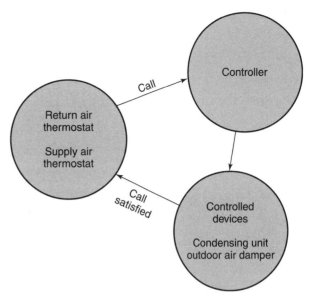

Figure 17.8 Control loop.

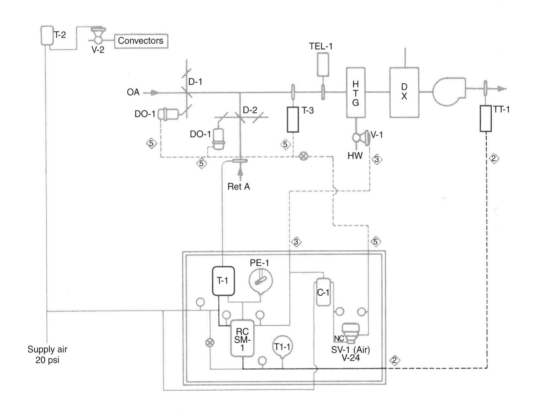

Legend

Ret A:	Return air	T-2:	Room thermostat
OA:	Outdoor air	T1-1:	Discharge air thermometer
C-1:	Minimum position controller	V-1:	Hot water valve
DO-1:	Outdoor air damper controller	V-2:	Room hot water valve
DO-2:	Return air damper controller		(convector)
D-1:	Outdoor air damper	TT-1:	Discharge air thermostat
D-2:	Return air damper	SV-1:	Air solenoid valve (outside
PE-1:	Cooling controller		air)
T-3:	Mixed air thermostat	TEL-1:	Freeze protector
RSCM-1:	Receiving controller	- - - - -	Air lines field installed
T-1:	Return air thermostat	———	Supply air lines

Figure 17.9 Pneumatic control system.

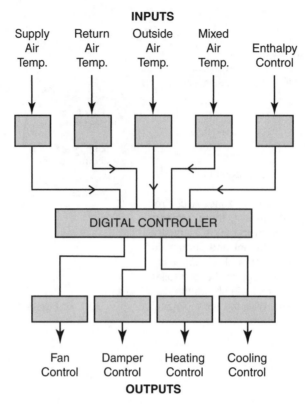

INPUTS

Figure 17.10 Inputs and outputs of a direct digital control system.

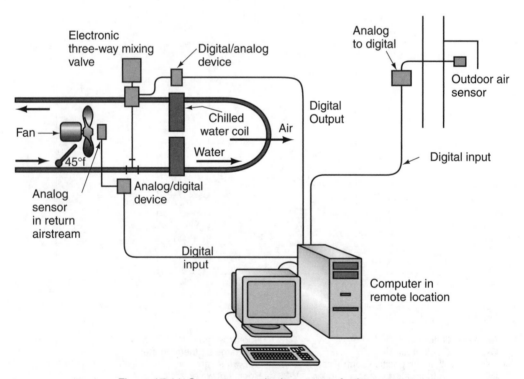

Figure 17.11 Computer monitoring system for large structure.

advancements in electronic industry, electronic controls have become the most popular method of control in commercial and industrial applications. Direct digital control uses a microprocessor that acts as a digital controller.

(Figure 17.10) shows the inputs and outputs of a direct digital control system. (Figure 17.11) shows a computer control and monitoring system for a large structure or complex?

REVIEW TEST

Name: _____	Date: _____	Grade: ____

Mark the following questions with a T for True or an F for false.

_____ 1. Most commercial and industrial systems use some method of capacity control because the load in a zone stays constant.

_____ 2. An across-the-line starter will energize the compressor with one contactor.

_____ 3. The part-winding starting system on a large commercial and industrial compressor uses 2 contactors.

_____ 4. In many large commercial and industrial air-conditioning systems, the blower motor operates constantly except when cut off by a time clock or for safety reasons.

_____ 5. The control system uses line voltage in most freestanding packaged units used in light commercial applications.

_____ 6. A pneumatic control system uses hydraulic fluid to operate the control system.

_____ 7. An electromechanical control system uses air pressure to operate the control system.

_____ 8. A water-cooled condenser unit with a cooling tower must have an interlock to ensure that the cooling tower fan is operating.

_____ 9. If a water chiller operates without the chilled water pump running, there is a possibility of the chiller freezing and rupturing.

_____ 10. An anti-short-cycling device installed in the compressor contactor coil circuit prevents the compressor from rapid starting and stopping.

_____ 11. A control loop has input signals that are sent to a controller that determines the correct outputs to maintain the structure temperature.

_____ 12. The pneumatic control system must have a constant supply of air at 75 psig that is clean and dry.

_____ 13. A pneumatic actuator could open or close a supply air damper or hot water valve.

_____ 14. The basic input to the receiver controller in (Figure 17.9) is T-1.

_____ 15. The direct digital control takes an analog signal and converts it to a digital signal.

_____ 16. A direct digital control operates on 24 volts AC.

_____ 17. A direct digital control system signal will allow a controlled device to modulate between full open and full close.

_____ 18. On a commercial and industrial condensing unit, the discharge pressure would be maintained by stopping the compressor.

_____ 19. A variable-speed fan motor could not be used on a direct digital control system.

_____ 20. An air line supplying a pneumatic control system must be a minimum of 1 ½ inches in diameter in order to supply the correct amount of air.

CHAPTER 18

Troubleshooting Refrigeration and Air-Conditioning Control Circuitry and Systems

Chapter Overview

The servicing of any air-conditioning or heating system is important because sooner or later the equipment will malfunction and need attention. Service technicians must develop a systematic approach to diagnosing faulty equipment and control circuits and making the necessary repairs. A good beginning for any service technician is to be able to understand and diagnose all the components of the system. Service technicians should use the many tools that are available. No matter how large or small an air-conditioning, heating, or refrigeration system is, the technician can break the control system into basic circuits and then check the circuits on an individual basis. This is much easier than attempting to treat the total diagram as one large circuit. By using a systematic approach to troubleshooting control systems and electrical components, service technicians can accomplish any troubleshooting task.

The technician must develop a systematic approach to troubleshooting but should also use the many tools that are available from manufacturers. The technician will soon find out that the most important tool in troubleshooting is a good electrical meter. Another valuable tool that a technician can use is the schematic diagram. It is attached to a cover in most equipment and is extremely useful when troubleshooting circuitry is necessary. Manufacturers furnish installation and troubleshooting information with each new piece of equipment that is sold. This material includes installation procedures and specifications that are needed to correctly size the circuit supplying the equipment, as well as refrigerant piping data. The troubleshooting part of this information should be left inside the control panel of the equipment for the service technician because of the valuable service manual. Troubleshooting charts are available from many manufacturers that provide a systematic approach to troubleshooting. A feature that is used on some new equipment is an electronic self-diagnosis, which helps the technician pinpoint the problem. Hopscotching is another tool that the technician can use to test circuits. Technicians should use the many troubleshooting tools that are available to them.

The technician must use a certain amount of common sense and observational skills to become a good troubleshooter. Many systems faults can be determined by a visual inspection of the equipment or control panels. Some common items that can be checked easily by visual inspection are shorted capacitors, bad contacts on a contactor and relay coils, burned or broken electrical connections, dirty condenser coils, and many others. There are many procedures used for troubleshooting refrigeration, heating, or air-conditioning equipment. A good technician will develop his or her own procedure and use it. One procedure for troubleshooting any air-conditioning, heating, or refrigeration system is the following:

1. Check the power source to the equipment.
2. If no power is available, correct the problem.
3. Check the control voltage.
4. If there is no control voltage, correct the problem.
5. Make a thorough check of the system to determine which load (or loads) is not working.
6. Locate the loads on the wiring diagram that are not operating properly and begin checking these circuits.
7. Each device in the suspected circuit is a possible cause of the problem and should be checked until the faulty device is found. Check the affected circuits.
8. It may be necessary to move to other circuits to check the coils of some of the devices.
9. Never overlook the obvious.

Remember that service work is a necessity in this industry. No customer wants to purchase an air-conditioning and heating system from a company if he or she cannot get the proper service or maintenance. Service personnel always reflect on the employer when they are sent out on a service call. They should always try to present themselves in a manner that will sell themselves as well as their employer. The appearance and attitude of a service technician are just as important as the ability to diagnose and correct the system problem.

Key Terms

Current-sensing lockout relay
Electronic self-diagnostic
feature

Fault isolation diagram
Hopscotching

Installation and service
instructions
Troubleshooting tree

REVIEW TEST

Name:_____   Date: _____   Grade: _____

1. If the resistance reading of a 3-horsepower hermetic compressor motor is 220 ohms between one of the compressor motor terminals and the compressor housing, the condition of the motor is _____.
   a. shorted
   b. open
   c. grounded
   d. good

2. A service technician reads 230 volts across a set of contacts. The contacts are _____.
   a. open
   b. closed

3. If a compressor is extremely hot to touch and is not operating, but line voltage is available to the compressor, which of the following is a possible cause?
   a. motor has bad bearings
   b. bad running capacitor
   c. bad start winding
   d. all of the above

4. Which of the following is a procedure used when checking electrical circuits?
   a. hopscotching
   b. double-checking
   c. ampere check
   d. all of the above

5. If the technician moves the fan switch on a thermostat to the "on" or "continuous" position and the fan motor runs, which of the following conditions has the technician confirmed if it is a split system?
   a. The transformer is good and power is being supplied to the indoor unit.
   b. The transformer is good.
   c. The thermostat is good.
   d. The power is being supplied to the outside unit.

6. It is essential that service technicians be able to read schematic diagrams.
   a. true
   b. false

7. A compressor and condenser fan motor are connected in parallel to a contactor. The compressor is operating, but the condenser fan motor is not. There are no broken connections. Which of the following components could be faulty?
   a. contactor
   b. condenser fan motor
   c. transformer
   d. thermostat

8. If a service technician reads the correct voltage to a contactor coil and the contactor is not closing, what is the problem?
   a. thermostat
   b. transformer
   c. contacts of contactor
   d. contactor coil

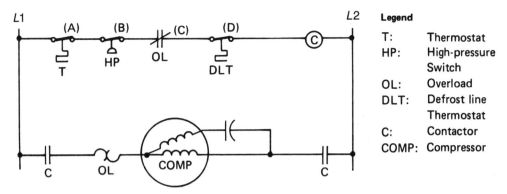

Figure 18.1 Schematic diagram for Questions 11 and 12.

9. **A technician reads 50 volts across a closed set of contactor contacts. What does this indicate?**
   a. good contactor coil
   b. bad contactor contacts
   c. bad transformer
   b. none of the above

10. **Service technicians rarely locate problems by visually inspecting the unit.**
    a. true
    b. false

11. **In Figure 18.1, if line voltage is at the _L2_ side of the contactor coil, and at A but is not at B, _____.**
    a. the thermostat is open
    b. the HP is open
    c. the OL is open
    d. the DLT is open

12. **In Figure 18.1, if the line voltage is available to the load side of the contactor, but is not available to the compressor, what could the problem be?**
    a. bad contactor
    b. bad compressor
    c. bad overload element
    d. bad overload contacts

13. **Which of the following problems might a technician locate by visually inspecting a condensing unit?**
    a. bad fuse
    b. bad compressor
    c. open manual reset high-pressure switch
    d. bad thermostat

14. **What would the results be if two safety controls were connected in parallel?**
    a. If one opened the circuit would be de-energized.
    b. Both would have to open to de-energize the circuit.
    c. A correct method of installing safety controls.
    d. none of the above

15. **Which of the following could be a problem if the technician was not getting voltage to a condensing unit?**
    a. fuse
    b. circuit breaker
    c. broken wire
    d. all of the above

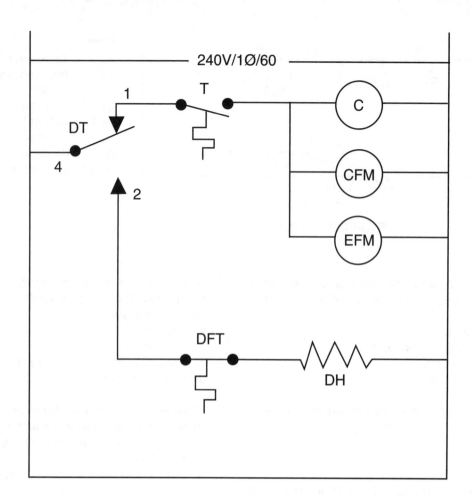

240V/1Ø/60

Legend

DT: Defrost timer
DTM: Defrost timer motor
T: Thermostat
C: Compressor
CFM: Condenser fan motor
EFM: Evaporator fan motor
DFT: Defrost thermostat
DH: Defrost heater

Figure 18.2 Schematic diagram for Questions 16 and 17.

Answer Questions 16 and 17 based on Figure 18.2.

16. **If the DH is not operating, which of the following combinations of electrical devices should be checked?**
   a. DT contacts 4 & 2, DTM, DFT, and DH
   b. DT contacts 4 & 2, DTM, and DH
   c. DT contacts 4 & 2, DFT, and DH
   d. DTM and DH

17. **The compressor will not start, the DT contacts 4 to 1 are closed, and 230 volts are being supplied to the unit—what is the cause of the problem?**
   a. DT
   b. T
   c. DFT
   d. CFM

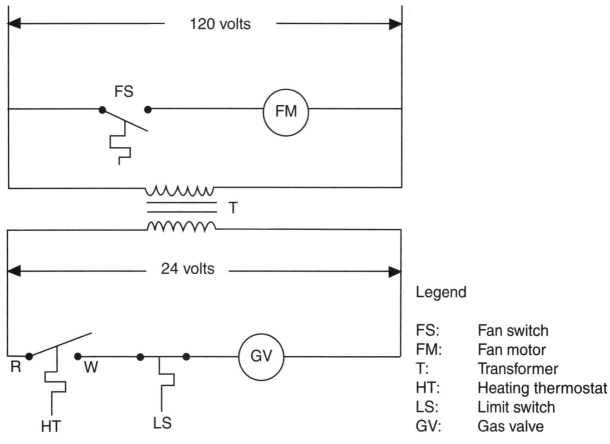

120 volts

FS

FM

T

24 volts

R

W

GV

HT

LS

Legend

FS: Fan switch
FM: Fan motor
T: Transformer
HT: Heating thermostat
LS: Limit switch
GV: Gas valve

Figure 18.3 Schematic diagram for Questions 18, 19, and 20.

Answer Questions 18, 19, and 20 based on Figure 18.3.

18. The furnace is being supplied with 120 volts, the HT and LS are closed, and the GV is good but will not energize—what is the problem with the furnace?
   a. FS
   b. FM
   c. T
   d. none of the above

19. The GV is energized, the gas is igniting in the combustion chamber, and the combustion chamber temperature is rising. Which of the following electrical devices could be faulty?
   a. FS
   b. FM
   c. LS
   d. all of the above

20. The temperature of the furnace combustion chamber is extremely hot and the fan motor is operating. What is the probable cause of the malfunction?
   a. HT
   b. LS
   c. GV
   d. T

Troubleshooting Refrigeration, Heating, or Air-Conditioning Systems

Name: _____	Date: _____	Grade: ___

Comments:

Objectives: Upon completion of this lab, you should be able to correctly troubleshoot basic air-conditioning problems assigned by your instructor.

Introduction: One of the most important jobs of air-conditioning technicians is to be able to locate and repair a system problem and return the air-conditioning system to normal operation. Approximately 80% of the problems in air-conditioning and heating systems will be electrical. Service technicians must be able to use electrical meters and read schematic diagrams in order to troubleshoot air-conditioning and heating systems.

Text Reference: Chapter 18

Tools and Materials: The following materials and equipment will be needed to complete this lab exercise.
  Operating heating and air-conditioning systems
  Electrical meters
  Basic electrical handtools

Safety Precautions: Make certain that the electrical source is disconnected when making electrical connections. In addition:
  • Make sure all connections are tight.
  • Make sure no bare conductors are touching metal surfaces except the grounding conductor.
  • Make sure the correct voltage is being supplied to the unit.
  • Make sure body parts do not come in contact with live electrical conductors.
  • Keep hands and materials away from moving parts.
  • Make sure all covers on the equipment are replaced.

LABORATORY SEQUENCE (mark each box upon completion of task)

A. Troubleshooting Problems

☐ 1. You answer a complaint of "no cooling." The condenser fan motor and evaporator fan are operating. You check the schematic diagram (see Figure 18.4) and decide that the problem is

_____.
   a. an open internal overload in the compressor
   b. blown fuses in the disconnect
   c. a bad transformer
   d. a bad thermostat

☐ 2. You answer a complaint of "not enough cooling." All unit components are operating, but after about five minutes, the compressor cuts out. You check the pressure of the refrigeration system and find that the discharge pressure is 160 psig and the suction pressure is 20 psig. The refrigerant in the system is R-22. You check the schematic diagram (see Figure 18.4) and decide that the problem is _____.
   a. a bad transformer
   b. a bad thermostat
   c. an open low-pressure switch
   d. an open high-pressure switch

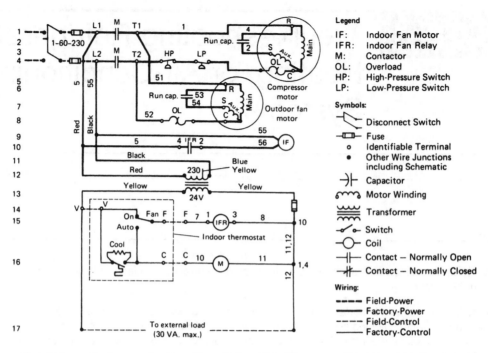

Figure 18.4 Schematic diagram for Questions 1, 2, and 3. *(Courtesy of York International Corporation)*

3. You answer a complaint of "no cooling." The compressor and condenser fan motor are operating, but the indoor fan motor is not (refer to Figure 18.4). The probable cause is _____.
   a. a bad indoor fan motor
   b. a bad transformer
   c. a bad indoor fan relay
   d. both a and c

4. You answer a complaint of "no heating." The compressor and outdoor fan motor are not operating but are good. The indoor fan is operating properly. After checking the schematic diagram (see Figures 18.5 and 18.6), you determine that the problem is a bad _____.
   a. transformer
   b. thermostat
   c. MS contactor
   d. supplementary heater

5. You answer a complaint of "insufficient heating." The compressor, the condenser fan motor and indoor fan motor are operating, and there is a heavy coating of ice on the outdoor unit coil. The defrost thermostat is closed. (See Figure 18.5 for the schematic diagram.) The probable cause is _____.
   a. a bad defrost timer
   b. a bad transformer
   c. a bad defrost relay
   d. both a and c

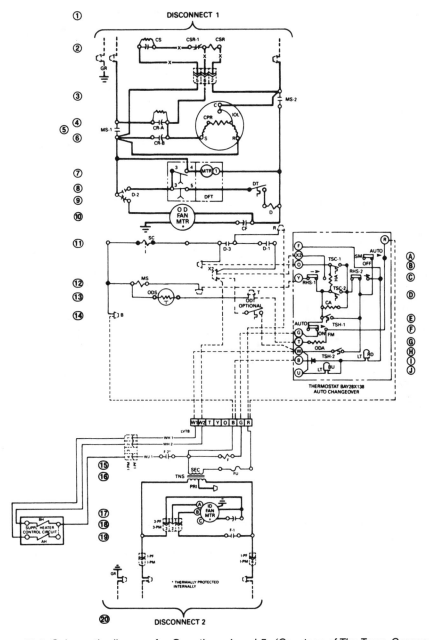

Figure 18.5 Schematic diagram for Questions 4 and 5. *(Courtesy of The Trane Company)*

Legend

AH:	Supplementary Heat Contactor	(19)	LT:	Light	
BH:	Supplementary Heat Contactor	(18)	LVTB:	Low-Voltage Terminal Board	
CA:	Cooling Anticipator		MS:	Compressor Motor Contactor	(5, 3, & 12)
CR:	Run Capacitor	(4 & 6)	MTR:	Motor	
CPR:	Compressor		ODA:	Outdoor Temperature Anticipator	(G)
D:	Defrost Relay	(9)	ODS:	Outdoor Temperature Sensor	(13)
DFT:	Defrost Timer	(7)	ODT:	Outdoor Thermostat	
DT:	Defrost Termination Thermostat	(8)	RHS:	Resistance Heat Switch	(C)
F:	Indoor Fan Relay	(15)	SC:	Switchover Valve Solenoid	(11)
FM:	Manual Fan Switch	(F)	SM:	System Switch	(A)
HA:	Heating Anticipator		TNS:	Transformer	(16)
HTR:	Heater		TSC:	Cooling Thermostat	(B & D)
IOL:	Internal Overload Protection		TSH:	Heating Thermostat	(E & H)

Figure 18.6 Legend of Figure 18.5.

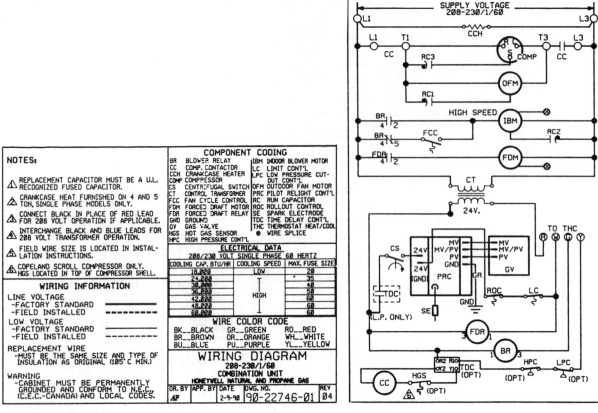

Figure 18.7 Schematic diagram for Questions 6 and 7. *(Courtesy of Rheem Air Conditioning Division, Fort Smith, AR)*

6. You answer a complaint of "no heating." The combustion chamber and blower section is extremely hot and the LC is open. The blower motor is good and there is no restriction in the supply air distribution system. The blower relay contacts and coil are good. The schematic diagram of the unit is shown in Figure 18.7. The probable cause is a bad _____.
   a. gas valve
   b. pilot relight control
   c. flame rollout switch
   d. fan cycle control

7. You answer a complaint of "no heating." The pilot ignites, but the main burner does not. You check, and 24 volts are available to the PRC but no voltage is available to the GV. The schematic diagram is shown in Figure 18.7. What is the probable cause?
   a. gas valve
   b. CS
   c. ROC
   d. PRC

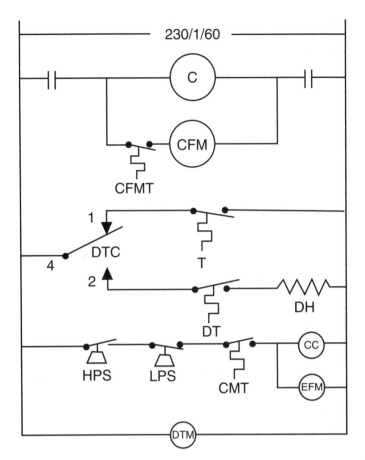

Legend

C: Compressor
CC: Compressor contactor
CFMT: Condenser fan motor thermostat
CFM: Condenser fan motor
DT: Defrost timer motor
DTC: Defrost timer contacts
T: Thermostat
LLS: Liquid line solenoid
DH: Defrost heater
HPS: High-pressure switch
LPS: Low-pressure switch
CMT: Compressor motor thermostat
EFM: Evaporator fan motor

Figure 18.8 Schematic diagram for Questions 8, 9, and 10.

8. You answer a complaint of rising temperature in a walk-in freezer. The evaporator is completely covered with frost. The schematic of the unit is shown in Figure 18.8. The unit is being supplied the correct voltage. The DTC contacts are good, and the DTM and the DH are good. The probable cause of the problem is _____.
   a. LLS
   b. open DT
   c. closed DT
   d. CC

9. You answer a complaint of a walk-in freezer not cycling off properly, and the temperature of the freezer is −55 °F. The schematic of the unit is shown in Figure 18.8. The thermostat is set at 0 °F. The probable cause of the problem is _____.
   a. DTC
   b. LPS
   c. CMT
   d. none of the above

10. You answer a complaint of a walk-in-freezer temperature of 40 °F. The schematic of the unit is shown in Figure 18.8. You discover that the discharge pressure reaches 500 psig when the HPS opens, stopping the compressor. The condenser fan motor is good. The probable cause is

   _____.
   a. CFMT
   b. LLS
   c. DH
   d. CMT

B. Troubleshooting Heating, Air-Conditioning, and Refrigeration Systems

☐   1. Your instructor will assign five systems for you to troubleshoot.

☐   2. Troubleshoot and record in the following table the problems found with the five systems.

Unit	System Problems
#1	_____
#2	_____
#3	_____
#4	_____
#5	_____

☐   3. Have your instructor check your work.

MAINTENANCE OF WORK STATION AND TOOLS: Clean and return all tools to their proper location(s). Replace all equipment covers. Clean up the work area.

SUMMARY STATEMENT: Briefly explain the procedure that you used for troubleshooting the five systems.

Questions

1. If no part of a unit is operating, what is the first check that a service technician should make?

2. What is hopscotching?

3. When a service technician arrives on the job, what are the steps that should be taken first?

4. A hermetic compressor in a small residential condensing unit is not operating, but voltage is available to the compressor terminals. What is the probable cause?

5. What would be some possible causes of a compressor motor humming when the contactor is closed?

6. What safety control would be likely to open if the unit had an extremely dirty condenser coil?

7. What are some common problems that occur with low-voltage thermostats?

8. What are some common causes of contactor failures?

9. Why are electric meters important to the service technician?

10. Why are schematic wiring diagrams important to the service technician?